Toxic Taste

How Processed Foods Are Slowly Killing You

Douglas B Sims, PhD

Toxic Taste

Douglas B Sims, PhD

For more information, or to book an event, contact:
dsims@simsassociates.net

Book design by DB Sims
Cover picture: iStock

ISBN – Paperback: 978-1-966739-03-6
ISBN – eBook: 978-1-966739-02-9

First Edition: February 2025

Toxic Taste

Table of Contents

Acknowledgements

I am profoundly grateful to my wife, whose steadfast support, wisdom, and love have been my anchor and inspiration. The journey we've shared over the past 34 years has enriched every chapter of my life and this project. Your encouragement has lifted me through each challenge, and your insights have helped shape my perspective in ways that are reflected on every page.

To our children, thank you for filling our lives with joy and growth, teaching us both the rewards and trials of parenthood. Watching you become who you are has been one of my life's greatest privileges, filling me with pride and offering lessons that influence my work and my worldview.

To our family, thank you for your unwavering support. Your presence has been a source of strength, and your companionship has been invaluable in both my personal journey and in completing this project.

To my friends and colleagues—especially those in renewable energy professions and political science—I extend my deepest gratitude. Your insights, expertise, and perspectives have challenged and enriched my thinking. Engaging in debates and discussions with you has added a depth and authenticity to this book that I could not have achieved alone.

Lastly, to the many individuals I have had the privilege of observing and interacting with professionally, thank you for sharing your experiences. Your stories have provided invaluable insights, adding real-world understanding that resonates throughout these pages.

Toxic Taste

Forward

In recent decades, the way we eat has undergone a profound transformation. The rise of processed foods, once a symbol of convenience and innovation, has dramatically reshaped global diets and impacted the health of society in ways we are only beginning to fully understand. This book explores the deep connections between processed foods and public health, offering a compelling examination of how the foods we consume affect not only our individual well-being but also the broader health of our communities.

As industrialization and globalization have made processed foods more accessible, they have also introduced new challenges. These foods, often loaded with added sugars, unhealthy fats, and artificial additives, have contributed to the rise of chronic diseases such as obesity, diabetes, and heart disease. The effects of these dietary patterns are particularly pronounced in vulnerable populations, where poor access to fresh, whole foods and limited nutritional education exacerbate health inequalities. This book sheds light on these pressing issues, connecting the dots between convenience-driven food choices and their far-reaching consequences for public health.

The relationship between processed foods and health is complex and multifaceted. While processed foods provide an immediate solution to the demands of busy modern lives—offering quick, affordable, and shelf-stable options—the long-term impact on our health and the environment has been profound. The growing body of research on the adverse effects of ultra-processed foods, particularly their role in chronic disease and environmental degradation, underscores the urgency of reevaluating our food systems. At the heart of this evaluation is the recognition that the food industry, governments, and consumers all play pivotal roles in shaping the future of food.

Through the lens of science, policy, and personal responsibility, this book offers not only a critique of the modern food landscape but also a hopeful vision for the future. It presents a call to action for all of us to reassess our food choices, support healthier, sustainable practices, and advocate for policies that make nutritious foods accessible to everyone. By reimagining our food systems—where convenience and health can coexist—we can create a future where the health of our bodies, our communities, and our planet are aligned.

As you read this book, you will be invited to consider your role in the food system and the impact your choices have, both personally and societally. We have the power to shape the future of food. With awareness, education, and collective action, we can move toward a world where the benefits of processed foods are balanced with the nourishment of whole, unprocessed options—creating a healthier, more sustainable world for generations to come.

Chapter 1

The Science on Your Plate

In today's fast-paced world, processed foods dominate our diets, offering convenience and affordability to meet the demands of modern life. From the brightly packaged snacks lining supermarket shelves to ready-to-eat meals that save precious time, these products have become an inescapable part of the global food system. While their rise has been driven by technological advancements and economic incentives, processed foods come with significant trade-offs. Behind their appeal lies a complex web of chemical transformations that enhance flavor, texture, and shelf life but often at the cost of nutritional quality. This book explores the science behind processed foods, delving into the delicate balance between their promises of convenience and the growing concerns about their impact on health, society, and the environment. Through this journey, we aim to demystify the chemistry of modern food and empower readers to make more informed choices.

Why Processed Foods Dominate Modern Diets

Processed foods have become a cornerstone of modern diets, shaped by the demands of convenience, affordability, and accessibility. In a world where schedules grow busier and time feels increasingly scarce, many people turn to ready-to-eat meals, frozen entrees, and packaged snacks as quick solutions for their nutritional needs. These foods require little to no preparation, aligning perfectly with the lifestyles of workers, students, and families juggling multiple responsibilities (Monteiro et al., 2018). The industrialization of food production plays a pivotal role in this reality. Advances in processing and preservation technologies enable the mass production, distribution, and storage of food, ensuring a constant supply of affordable and long-lasting options.

Economic factors further drive the dominance of processed foods. Government subsidies for staple crops like corn, soy, and wheat—essential ingredients in many processed products—reduce production costs and make these items cheaper than fresh alternatives (FAO, 2021). For low-income families, processed foods often represent the most cost-effective way to feed their households, even if they come at the expense of nutritional quality. These affordability dynamics perpetuate reliance on products like cereals, snack bars, and instant noodles, which are readily accessible in supermarkets, convenience stores, and vending machines worldwide.

Marketing strategies also amplify the popularity of processed foods. Food companies invest billions of dollars annually in advertising campaigns that highlight convenience, affordability, and indulgent flavors, while downplaying nutritional concerns (Nestle, 2013). These campaigns often target specific demographics, such as working parents, teenagers, and children, with messaging that aligns processed foods with pleasure, ease, and modern living. The strategic use of vibrant packaging, health claims, and endorsements further cements their appeal, influencing purchasing decisions and shaping eating habits.

Cultural factors also play a role in the proliferation of processed foods. In many societies, traditional cooking skills and home-prepared meals are declining due to shifts in family dynamics, urbanization, and the rise of dual-income households. Processed foods fill the gap left by these changes, offering a quick solution for meals that can be prepared with minimal effort. As these products dominate grocery stores, fast-food outlets, and even school cafeterias, they become normalized as staple components of daily diets.

While processed foods offer convenience and affordability, they also contribute to a dependency on products that prioritize taste and shelf stability over nutritional value. Their widespread availability has reshaped global diets, often leading to increased consumption of added sugars, unhealthy fats, and sodium, while displacing fresh, whole foods. This dependency underscores the need to critically examine the factors driving the proliferation of processed foods and their broader implications for health, culture, and the environment. By understanding these dynamics, individuals and policymakers alike can begin to address the challenges posed by an overreliance on processed foods and work toward a healthier, more sustainable food system.

An Overview of the Chemical Transformations in Food Production

Food processing relies on a series of sophisticated chemical transformations to produce items that are safe, palatable, and convenient for consumers. These transformations enhance taste, texture, longevity, and nutritional content, often starting with raw agricultural products such as grains, fruits, vegetables, and oils. These raw materials are refined and altered to create base ingredients like flours, starches, oils, and syrups, which serve as the building blocks of many processed foods (Institute of Food Technologists [IFT], 2018). The chemical engineering involved in these processes has revolutionized the food industry, enabling the creation of products that meet the demands of modern lifestyles.

Additives play a crucial role in processed foods, performing functions that are integral to their appeal and shelf stability. Preservatives, such as sodium benzoate and potassium sorbate, prevent microbial growth and spoilage, allowing products to remain safe for consumption over extended periods (Moss, 2013). These substances are essential for mass production and distribution, especially for products that must travel long distances before reaching consumers. Emulsifiers, such as lecithin, are another common additive that prevents ingredients like oil and water from separating, ensuring consistent textures in items like mayonnaise, salad dressings, and ice creams. Stabilizers and thickeners, such as guar gum and xanthan gum, add viscosity and improve mouthfeel, enhancing the sensory experience of processed foods.

Flavor enhancers are another key category of chemical additives. Compounds like monosodium glutamate (MSG) amplify savory flavors, making processed foods more appealing. Artificial and natural flavoring agents mimic or intensify the taste of fresh ingredients, compensating for the loss of flavor during processing. Similarly, artificial colors improve the visual appeal of foods, reinforcing consumer perceptions of freshness and quality. These additives are meticulously formulated to ensure consistency and uniformity, which are hallmarks of processed food products.

One of the most significant chemical transformations in food production is hydrogenation, a process that converts liquid unsaturated fats into solid or semi-solid saturated fats to improve shelf stability and texture. Partially hydrogenated oils, which became popular in the mid-20th century, were initially lauded for their ability to extend the shelf life of baked goods, margarine, and snack foods. However, the hydrogenation process produces trans fats, now widely recognized for their role in raising bad cholesterol (LDL) and lowering good cholesterol (HDL), contributing to heart disease and other health issues (Willett & Stampfer, 2013). As a result, regulatory agencies worldwide have moved to restrict or ban trans fats in recent years.

Sweeteners also play a prominent role in food processing. High-fructose corn syrup (HFCS), derived from cornstarch, emerged as a cost-effective alternative to cane sugar in the 1970s. Its use revolutionized the production of sweetened beverages, baked goods, and condiments. However, HFCS has been implicated in the rise of obesity, diabetes, and other metabolic disorders due to its high caloric content and its effect on insulin resistance (Bray et al., 2004). Artificial sweeteners, such as aspartame and sucralose, have been developed as low-calorie alternatives, but their long-term health effects remain a subject of ongoing research and debate.

Enrichment and fortification are additional chemical processes used to enhance the nutritional profile of processed foods. Vitamins and minerals are often added to products like cereals, bread, and dairy to address nutrient deficiencies in the population. For instance, the fortification of flour with folic acid has significantly reduced the incidence of neural tube defects in newborns. While these processes have clear public health benefits, they also raise questions about the bioavailability and efficacy of synthetic nutrients compared to those naturally present in whole foods.

Food preservation techniques, such as canning, freezing, and dehydration, involve chemical and physical transformations that extend the shelf life of perishable items. These methods often rely on chemical preservatives or inert gases to prevent spoilage and oxidation. While they make it possible to store and transport food safely over long distances, they may also reduce the nutritional value of the final product. For example, heat treatments used in canning can degrade heat-sensitive vitamins like vitamin C and some B vitamins.

While these chemical transformations have enabled the mass production of affordable and convenient foods, they also raise important questions about their long-term impact on health and well-being. The prevalence of additives, preservatives, and chemically modified ingredients in processed foods has prompted scrutiny from health professionals and researchers, who point to rising rates of obesity, heart disease, and diabetes as potential consequences of overconsumption. As these transformations continue to shape the global food supply, a deeper understanding of their implications is critical for informed consumer choices and the development of healthier, more sustainable food systems.

The Promise and Peril of Processed Foods: Convenience vs. Consequences

Processed foods undeniably bring significant benefits, making them a cornerstone of modern diets. One of their greatest advantages is the ability to reduce food waste. By extending shelf life and improving transportability, processed foods ensure that food remains safe and consumable over long periods, even when distributed across vast distances. This is particularly crucial in global food supply chains, where fresh food might otherwise spoil before reaching consumers. The convenience of processed foods cannot be overstated, as they require minimal preparation and cater to the fast-paced lifestyles of modern society.

Affordability is another key strength of processed foods. By utilizing mass production techniques and subsidized staple crops like corn, soy, and wheat, food manufacturers can produce inexpensive products that provide accessible calories. For individuals living in food deserts— regions where fresh produce is scarce—or low-income communities, processed foods often serve as a lifeline, offering essential calories and nutrients when fresh alternatives are unavailable (Drewnowski & Darmon, 2005). Additionally, fortification of processed foods has played a vital role in addressing nutrient deficiencies on a global scale. Products such as fortified cereals, iodized salt, and vitamin D-enriched milk have significantly improved public health by preventing conditions like rickets, goiter, and neural tube defects (FAO, 2021).

However, the over-reliance on processed foods comes with significant consequences, both for individual health and the environment. A growing body of research links the consumption of ultra-processed foods to a range of chronic health conditions. These products are often calorie-dense but nutrient-poor, packed with added sugars, unhealthy fats, and excessive sodium. This imbalance contributes to widespread health issues such as obesity, type 2 diabetes, hypertension, and cardiovascular diseases (Monteiro et al., 2018). The long-term health risks associated with chemical additives, artificial colors, and preservatives commonly found in processed foods are increasingly scrutinized. While many additives are deemed safe at regulated levels, concerns about cumulative exposure and their potential impact on gut health, metabolic function, and allergies persist (Moss, 2013).

Beyond individual health, the environmental costs of industrial food production are staggering. The reliance on monoculture farming for staple ingredients leads to soil degradation, reduced biodiversity, and increased vulnerability to pests and diseases. This approach also contributes to deforestation and water overuse, exacerbating environmental challenges. Furthermore, the extensive use of single-use plastics for packaging processed foods generates enormous amounts of waste, much of which ends up in landfills or polluting oceans. These environmental impacts highlight the unsustainable nature of current food production systems (FAO, 2021).

The duality of processed foods—offering both promise and peril—requires a nuanced understanding of their role in society. On one hand, they address critical issues like food accessibility, affordability, and nutrient deficiencies. On the other, their widespread consumption exacerbates public health crises and environmental degradation. Moving forward, a balanced approach is necessary. This involves promoting the benefits of processed foods while mitigating their negative impacts through innovation, regulation, and education. By rethinking how we produce and consume food, we can work toward a system that prioritizes both individual health and environmental sustainability.

Goals of the Book: Demystify the Chemistry and Explore the Health and Societal Impacts

The aim of this book is to explore processed foods from multiple angles: scientific, health-related, and societal. By demystifying the chemical processes behind food production, readers can better understand the composition of the foods they consume. This knowledge empowers individuals to make informed dietary choices and challenge misleading marketing narratives (Nestle, 2013).

In addition, the book examines the health implications of processed foods, linking chemical additives and synthetic compounds to chronic diseases. It highlights the need for regulatory reform to ensure that food safety standards align with current scientific understanding (Bray et al., 2004). On a broader scale, the book explores the environmental and societal costs of industrial food production, emphasizing the urgent need for sustainable practices.

Ultimately, this book seeks to inspire change, whether through personal dietary adjustments, advocacy for better industry practices, or support for policies that prioritize public health and sustainability. By providing a comprehensive analysis of the science and impact of processed foods, it equips readers with the tools to navigate the complexities of modern food systems and envision a healthier, more sustainable future.

Processed foods have revolutionized the way we eat, offering unprecedented convenience and accessibility. However, their dominance has also introduced significant challenges, from health risks linked to poor nutrition and chemical additives to environmental and societal consequences that affect us all. Understanding the science behind processed foods is essential to navigating these complexities and making informed decisions about what we consume. As we move forward, balancing innovation with sustainability and health will be crucial to shaping a better food system—one that nourishes individuals, protects the planet, and prioritizes long-term well-being over short-term convenience. By exploring the chemistry, impact, and potential of processed foods, this book aims to inspire change at every level, from personal choices to global policies. Together, we can envision and create a future where the science on our plates truly serves the greater good.

Chapter 2

What Are Processed Foods?

Processed foods are a ubiquitous part of modern diets, ranging from simple canned vegetables to elaborate packaged snacks. Understanding what constitutes a processed food, the reasons for processing, and the ingredients involved is essential for evaluating their impact on health, society, and the environment. This chapter explores the definitions and categories of processed foods, the purposes behind food processing, and the key ingredients and additives that define these products.

Definitions and Categories of Processed Foods

The term "processed food" is broad and encompasses an extensive range of products that differ in how they are prepared, preserved, and modified. The extent of processing varies from minimal interventions to extensive industrial modifications. These variations determine not only the classification of the food but also its nutritional quality and potential health implications. The below will provide an extensive description of these three categories.

Minimally Processed Foods

Minimally processed foods are those that have undergone basic changes to enhance their shelf life, safety, or convenience while preserving most of their original nutritional content. These modifications are often limited to cleaning, cutting, freezing, drying, or pasteurization. Examples include washed and pre-packaged salads, frozen fruits and vegetables, and pasteurized milk. These foods are typically close to their natural state and are considered healthful options because they retain their essential vitamins, minerals, and other nutrients (Monteiro et al., 2018).

Freezing is a common preservation technique for minimally processed foods. It locks in nutrients and ensures availability year-round, even for seasonal produce. Similarly, pasteurization is used to eliminate harmful bacteria in milk and juice, making these products safe for consumption without significantly altering their nutritional value. These interventions improve accessibility and convenience while maintaining the health benefits of whole foods.

For many, minimally processed foods are a practical solution for incorporating more whole foods into their diets. For example, pre-washed spinach or frozen berries save time without compromising nutritional quality. These foods are often recommended as staples for health-conscious consumers because they strike a balance between convenience and nutritional integrity.

Processed Foods

Processed foods occupy the middle ground between minimally processed and ultra-processed items. These foods are altered to enhance flavor, texture, or shelf life, often through the addition of ingredients like salt, sugar, oil, or preservatives. Common examples include canned vegetables, cheese, bread, and smoked fish. While these foods often retain some of their original nutritional value, the inclusion of additional ingredients can affect their health profile.

For instance, canned vegetables retain their fiber and many vitamins but may also contain added sodium as a preservative. Similarly, bread provides essential carbohydrates but often includes refined flour and added sugars, which can diminish its nutritional quality compared to whole-grain alternatives. Cheese, while a good source of calcium and protein, may also be high in saturated fats and salt.

Processed foods are a significant part of many diets because of their convenience, affordability, and longer shelf life. However, their healthfulness largely depends on the extent of modifications and the quality of the added ingredients. While some processed foods can be part of a balanced diet, others, particularly those with high levels of added sugars, fats, or sodium, should be consumed in moderation.

Ultra-Processed Foods

Ultra-processed foods represent the most heavily modified category of processed foods. These are industrial formulations made primarily from refined ingredients and additives, often bearing little resemblance to their original source materials. Examples include sugary breakfast cereals, carbonated soft drinks, packaged cookies, instant noodles, and frozen ready meals. Ultra-processed foods are engineered to be hyper-palatable, combining flavors, textures, and colors that make them highly appealing to consumers (Monteiro et al., 2018).

The production of ultra-processed foods often involves the use of artificial flavors, colors, emulsifiers, and other additives that enhance sensory appeal and shelf stability. These foods are typically calorie-dense but nutrient-poor, offering high levels of sugar, unhealthy fats, and sodium while lacking fiber, vitamins, and essential minerals. Their low nutritional quality and high energy density contribute to overconsumption, making them a key factor in the global rise of diet-related chronic diseases such as obesity, type 2 diabetes, and cardiovascular diseases (Fardet, 2016).

Ultra-processed foods are convenient and affordable, making them accessible to a broad range of consumers. However, their widespread consumption has significant public health implications. Research has linked diets high in ultra-processed foods to negative health outcomes, including increased risk of hypertension, metabolic disorders, and certain types of cancer. Additionally, their addictive-like qualities, stemming from their hyper-palatable nature, can lead to overeating and displacement of healthier food options.

The NOVA Food Classification System

To better understand the role of processing in food systems, researchers led by Monteiro developed the NOVA food classification system. This framework categorizes foods into four groups based on the extent and purpose of their processing:

Group 1: Unprocessed or Minimally Processed Foods

Unprocessed or minimally processed foods are the foundation of a healthy diet. These items are in their natural state or have undergone minimal modifications to enhance safety, shelf life, or convenience. Examples include fresh fruits and vegetables, whole grains, eggs, fresh fish, raw nuts, and pasteurized milk. Processing methods in this category typically involve washing, cutting, freezing, or pasteurizing to preserve the natural characteristics of the food (Monteiro et al., 2018).

Minimally processed foods retain their full nutritional profile, including essential vitamins, minerals, fiber, and antioxidants. For example, frozen vegetables are flash-frozen shortly after harvest, preserving their nutrient content while extending shelf life. Similarly, pasteurization eliminates harmful pathogens in milk without significantly altering its nutritional composition.

These foods are considered the most healthful because they provide the nutrients necessary for growth, development, and disease prevention without added fats, sugars, or chemicals. Diets rich in Group 1 foods are associated with reduced risks of chronic diseases, including heart disease, diabetes, and obesity. Encouraging consumption of these foods is a cornerstone of public health initiatives worldwide.

Group 2: Processed Culinary Ingredients

Processed culinary ingredients are substances derived from Group 1 foods through basic processing methods. These include items like vegetable oils, sugar, butter, salt, and honey. While they are not typically consumed on their own, they are essential for cooking and preparing meals, adding flavor, texture, and preservation to dishes (Monteiro et al., 2018).

These ingredients often undergo processes such as refining, milling, pressing, or extraction. For example, vegetable oils are produced by pressing seeds or nuts, while sugar is extracted and refined from sugarcane or sugar beets. Salt, one of the oldest known preservatives, is harvested from seawater or mined from underground deposits.

While Group 2 ingredients enhance the flavor and utility of meals, excessive consumption can pose health risks. High intake of salt, sugar, and unhealthy fats is linked to hypertension, cardiovascular disease, and obesity. Public health guidelines often emphasize moderation in the use of these ingredients to balance flavor and healthfulness in meal preparation.

Group 3: Processed Foods

Processed foods represent a middle ground between minimally processed and ultra-processed items. These are products made by combining Group 1 and Group 2 foods with additional ingredients like salt, sugar, or oil. Examples include canned vegetables, cheeses, breads, and smoked fish. These foods are typically altered to enhance flavor, texture, or shelf life while retaining some of their original nutritional value.

Processing methods in this category include canning, fermenting, baking, or smoking. For instance, canned vegetables are preserved with brine or vinegar to prevent spoilage, while cheese is created through fermentation and the addition of salt for flavor and preservation. Bread, a staple in many diets, combines refined flour, water, yeast, and salt to produce a versatile and shelf-stable product.

While processed foods offer convenience and accessibility, their nutritional value can vary widely depending on the extent of processing and the quality of added ingredients. Canned vegetables, for example, retain much of their fiber and vitamins but may contain high levels of sodium. Similarly, bread made with whole grains is a healthful source of carbohydrates, but bread made with refined flour and added sugars may contribute to weight gain and blood sugar spikes.

Processed foods can play a role in balanced diets, but consumers are encouraged to choose options with minimal added fats, sugars, and salt to maximize health benefits.

Group 4: Ultra-Processed Foods

Ultra-processed foods are industrially formulated products designed to be hyper-palatable, convenient, and highly marketable. They are typically made from refined ingredients and synthetic additives, bearing little resemblance to their original food sources. Examples include sugary breakfast cereals, soft drinks, packaged snacks, instant noodles, and frozen ready meals (Monteiro et al., 2018).

The production of ultra-processed foods involves extensive modification, including the use of emulsifiers, artificial flavors, colors, stabilizers, and preservatives. These additives enhance sensory appeal and shelf life, making the products attractive to consumers. For example, emulsifiers prevent separation in products like salad dressings, while artificial flavors replicate the taste of natural ingredients at a fraction of the cost.

Ultra-processed foods are calorie-dense but nutrient-poor, often high in added sugars, unhealthy fats, and sodium while lacking fiber, protein, and essential vitamins and minerals. These characteristics contribute to overconsumption, as their hyper-palatable nature can bypass the body's natural satiety mechanisms, leading to overeating. Research has linked diets high in ultra-processed foods to increased risks of obesity, type 2 diabetes, hypertension, and certain cancers (Fardet, 2016).

Despite their health risks, ultra-processed foods dominate global markets due to their affordability, convenience, and aggressive marketing. They are especially prevalent in low-income and urban areas where access to fresh foods is limited. Efforts to reduce consumption of ultra-processed foods are a priority for public health advocates, who emphasize the importance of shifting dietary patterns toward minimally processed and whole foods.

The NOVA classification system highlights the role of food processing in shaping modern diets and health outcomes. It underscores the significant nutritional differences between minimally processed, processed, and ultra-processed foods. By providing a clear framework for categorizing foods, NOVA helps consumers, researchers, and policymakers understand the impact of processing on diet quality and develop strategies to promote healthier eating habits.

This framework also encourages greater transparency in food labeling and marketing, empowering consumers to make informed choices. By prioritizing Group 1 and 2 foods and limiting consumption of Group 4 items, individuals can improve their dietary patterns and reduce the risk of diet-related chronic diseases. The NOVA system serves as a valuable tool for navigating the complexities of modern food systems and fostering a more health-conscious approach to nutrition.

Understanding the distinctions between minimally processed, processed, and ultra-processed foods is essential for making informed dietary choices and addressing the health challenges associated with modern diets. Each category of food processing comes with distinct nutritional profiles, potential health benefits, and risks. While minimally processed foods generally align with dietary recommendations and provide essential nutrients, processed and ultra-processed foods vary widely in their health impacts, requiring consumers to navigate food choices carefully to prioritize long-term well-being.

Minimally processed foods, such as fresh fruits, vegetables, whole grains, and unprocessed proteins, are the foundation of a balanced diet. These foods are rich in essential vitamins, minerals, fiber, and antioxidants, which contribute to overall health and disease prevention. Their minimal alteration ensures that their nutritional integrity remains intact. By emphasizing minimally processed foods in their diets, individuals can align with public health guidelines that promote whole foods as a cornerstone of nutrition.

Processed foods, which include canned goods, cheeses, and breads, occupy a middle ground. While these foods often retain some of their original nutrients, the addition of ingredients like salt, sugar, and oil can alter their health profiles. For example, canned vegetables provide fiber and essential vitamins but may contain added sodium for preservation. Similarly, bread made with whole grains can be a nutritious source of carbohydrates, whereas refined versions with added sugars may contribute to weight gain and other health concerns. Moderation and careful selection of processed foods can help individuals balance convenience with nutritional value.

The dominance of ultra-processed foods in global markets has significant implications for public health. These products are engineered for convenience and affordability, making them appealing to consumers with limited time or resources. In many low-income households and regions with limited access to fresh produce, ultra-processed foods are often the only affordable option. However, their widespread consumption has a hidden cost: their poor nutritional quality and hyper-palatable nature contribute to overconsumption and diet-related chronic diseases, including obesity, type 2 diabetes, hypertension, and cardiovascular diseases (Monteiro et al., 2018).

Ultra-processed foods are typically calorie-dense but nutrient-poor, high in added sugars, unhealthy fats, and sodium. Their hyper-palatable nature is designed to appeal to taste preferences, often bypassing natural satiety mechanisms and leading to overeating. This overconsumption is a major driver of the global epidemic of non-communicable diseases, placing an increasing burden on healthcare systems worldwide. The affordability and aggressive marketing of these products exacerbate the issue, particularly among vulnerable populations.

Beyond health, the prevalence of ultra-processed foods raises concerns about the sustainability of food systems. The production of these products relies heavily on monoculture farming, which depletes soil health, reduces biodiversity, and increases reliance on chemical fertilizers and pesticides. Additionally, the extensive use of packaging, particularly single-use plastics, contributes to environmental degradation, generating significant waste that often ends up in landfills or oceans. These environmental impacts underscore the need for a more sustainable approach to food production and consumption.

The NOVA classification system provides a valuable framework for addressing these challenges. By categorizing foods based on their level of processing, it highlights the nutritional disparities between minimally processed, processed, and ultra-processed products. This framework serves as a tool for researchers, policymakers, and consumers to evaluate the healthfulness of modern diets and make more informed decisions. It also underscores the importance of reducing reliance on ultra-processed foods and prioritizing whole and minimally processed options.

For individuals, understanding these distinctions empowers better food choices. Reading food labels, recognizing added sugars and unhealthy fats, and prioritizing nutrient-dense options can help consumers improve their diets. For policymakers, the NOVA system offers insights into where regulatory efforts should focus, such as limiting the marketing of ultra-processed foods to children, improving labeling transparency, and incentivizing the production and consumption of minimally processed foods.

The implications for public health are profound. By shifting dietary patterns away from ultra-processed products and toward minimally processed and whole foods, society can address the root causes of many chronic diseases. This shift requires collective action from individuals, governments, and the food industry to create a food environment that supports healthier choices. Public education campaigns, subsidies for fresh produce, and stricter regulations on additives and marketing practices are critical steps in this process.

In conclusion, the distinctions between different levels of food processing are not merely academic—they have real-world implications for health, sustainability, and the future of food systems. By embracing the NOVA framework and prioritizing minimally processed foods, individuals and societies can take meaningful steps toward improving public health outcomes and creating a more sustainable and equitable food system. Recognizing the risks associated with ultra-processed foods and acting to reduce their prevalence is essential for a healthier future.

Why Processing is Used: Preservation, Flavor Enhancement, and Convenience

Food processing serves a multitude of purposes that respond to both consumer needs and the demands of the food industry. Its primary objectives are to preserve food, enhance flavor, and provide convenience, all of which have shaped modern food systems. However, these benefits come with trade-offs that warrant careful consideration, particularly in terms of nutritional quality and health implications.

Preservation

Preservation is one of the most fundamental and longstanding purposes of food processing. Humans have been preserving food for centuries, using methods such as salting, smoking, and drying to extend the life of perishable items. Modern preservation techniques, such as pasteurization, canning, and freezing, have dramatically improved the safety, availability, and longevity of food products. These methods not only reduce food waste but also play a critical role in ensuring food security.

Pasteurization, developed in the 19th century, is used to eliminate harmful pathogens in products like milk, juice, and eggs, making them safe for consumption while maintaining their nutritional value. Freezing, another widely used method, preserves fruits, vegetables, and meats by halting microbial activity and enzyme reactions that cause spoilage. This technique locks in nutrients, ensuring that foods remain nutritious even after months in storage. For example, frozen berries and vegetables retain their vitamins and minerals, making them a reliable alternative to fresh produce, particularly in regions with seasonal limitations (Institute of Food Technologists [IFT], 2018).

Canning involves heating food in airtight containers to kill bacteria and extend shelf life. Canned goods, such as beans, soups, and vegetables, provide a consistent source of nutrition in areas where fresh produce may be scarce or prohibitively expensive. This is especially important in regions with limited access to refrigeration or where supply chains are disrupted.

These preservation techniques have not only revolutionized food storage but also addressed critical issues such as food insecurity and global distribution. However, some methods, like canning, can lead to nutrient loss, particularly for heat-sensitive vitamins such as vitamin C and some B vitamins. Balancing preservation with nutrient retention remains an ongoing challenge in the food industry.

Flavor Enhancement

Flavor is a critical factor influencing consumer food choices, and food processing plays a central role in enhancing the taste of products. Additives such as salt, sugar, spices, and artificial flavorings are used to make foods more appealing and enjoyable. These enhancements cater to consumer preferences for specific flavors, such as sweetness or savoriness, and ensure consistency in taste across batches.

Salt has been used as a flavor enhancer and preservative for centuries, while sugar is commonly added to beverages, cereals, and baked goods to improve palatability. Sweeteners, such as high-fructose corn syrup, have become ubiquitous in processed foods due to their affordability and ability to amplify sweetness. However, excessive sugar consumption has been linked to obesity, diabetes, and other metabolic disorders, raising concerns about its widespread use (Bray et al., 2004).

Flavor enhancers like monosodium glutamate (MSG) intensify savory tastes, creating a "umami" flavor that is highly appealing to consumers. MSG is commonly used in processed snacks, soups, and ready-to-eat meals to improve their palatability (Moss, 2013). Similarly, artificial flavorings replicate natural tastes, allowing manufacturers to produce food with consistent flavor profiles even when natural ingredients are unavailable or cost-prohibitive.

While flavor enhancement improves the sensory appeal of processed foods, it can also encourage overconsumption. Foods engineered to be hyper-palatable often bypass the body's natural satiety signals, leading to overeating and contributing to diet-related health issues. The challenge for the food industry lies in creating products that are both flavorful and nutritionally balanced.

Convenience Factor

Convenience has become a cornerstone of food processing, particularly in urbanized societies where time constraints significantly influence dietary choices. Ready-to-eat meals, pre-packaged snacks, and instant beverages cater to the needs of busy individuals and families, providing quick and easy solutions for meal preparation.

Processing techniques such as pre-cooking, pre-portioning, and vacuum-sealing reduce the time and effort required to prepare meals. For example, pre-washed salad greens eliminate the need for cleaning and chopping, while frozen pizza offers a complete meal that can be heated and served within minutes. Instant beverages like coffee and powdered drinks provide quick refreshment without the need for extensive preparation.

Convenience foods are especially valuable in dual-income households, where both adults may lack the time or energy to cook meals from scratch. Similarly, they serve as a lifeline for students, workers, and others with unpredictable schedules. The ability to store, transport, and consume these foods with minimal effort has made them indispensable in modern lifestyles (Drewnowski & Almiron-Roig, 2010).

However, convenience often comes at the expense of nutritional quality. Many convenience foods are ultra-processed and high in added sugars, unhealthy fats, and sodium. The reliance on these products can lead to poor dietary habits and increase the risk of chronic diseases. Balancing convenience with healthfulness is a pressing challenge for both consumers and food manufacturers.

Balancing Benefits and Trade-Offs

The benefits of food processing—preservation, flavor enhancement, and convenience—are undeniable. These techniques have improved food safety, reduced waste, and transformed the way we eat. However, they also underscore significant trade-offs, particularly when processing involves the addition of unhealthy ingredients or the removal of beneficial nutrients.

For consumers, making informed choices about processed foods requires understanding the methods and ingredients used in their production. Opting for minimally processed options, such as frozen fruits and vegetables, can help balance convenience with nutritional quality. For manufacturers, the challenge lies in developing processing techniques and formulations that prioritize health without compromising flavor or shelf life.

As food systems continue to evolve, addressing these trade-offs will be essential for promoting healthier diets and creating a more sustainable and equitable food supply.

Key Ingredients and Additives in Processed Foods

Processed foods often rely on a wide array of ingredients and additives to achieve their desired taste, texture, and shelf life. These components are carefully formulated to meet consumer expectations and regulatory standards.

Preservatives are among the most common additives used in food processing. Compounds like sodium benzoate and potassium sorbate inhibit the growth of bacteria, mold, and yeast, ensuring the safety and longevity of food products. For example, sodium nitrite is used in cured meats like bacon and sausage to prevent spoilage and enhance flavor. While these additives play a critical role in food safety, some have raised concerns due to potential links to health issues, including cancer, when consumed in excess (Lahou et al., 2017).

Emulsifiers and stabilizers are used to maintain the consistency of processed foods. Emulsifiers, such as lecithin and mono- and diglycerides, prevent the separation of oil and water in products like salad dressings, mayonnaise, and ice creams. Stabilizers like carrageenan and guar gum add viscosity, ensuring a smooth texture in dairy products, soups, and sauces (IFT, 2018). These additives contribute to the sensory appeal of foods but have also been scrutinized for their potential effects on gut health.

Sweeteners are a defining feature of many processed foods. High-fructose corn syrup (HFCS) and artificial sweeteners like aspartame and sucralose are commonly used to replace or supplement traditional sugar. While these ingredients reduce costs and calorie content, their overuse has been linked to metabolic disorders, obesity, and other health concerns (Bray et al., 2004).

Artificial colors and flavors enhance the sensory experience of processed foods, making them visually appealing and flavorful. Synthetic dyes like Red 40 and Yellow 5 are widely used in candies, beverages, and baked goods, while artificial flavorings replicate or amplify natural tastes. Although approved for use by regulatory agencies, some artificial additives have been associated with behavioral and allergic reactions, prompting calls for more transparent labeling and natural alternatives (Moss, 2013).

Fortification and enrichment are used to improve the nutritional profile of processed foods. For example, bread and cereals are often fortified with iron and B vitamins, while milk is enriched with vitamin D. These practices have significantly reduced nutrient deficiencies in populations worldwide. However, there is ongoing debate about whether synthetic nutrients are as bioavailable as those naturally present in whole foods (FAO, 2021).

These ingredients and additives exemplify the complex chemistry behind processed foods. While they enhance safety, convenience, and sensory appeal, their potential health risks highlight the need for balanced consumption and regulatory oversight.

Processed foods encompass a diverse range of products, from minimally processed to highly engineered, ultra-processed items. They play an essential role in modern diets by providing preservation, flavor enhancement, and convenience. However, the ingredients and additives that define these foods also pose challenges to health and well-being, especially when consumed in excess. By understanding the categories of processed foods, the purposes behind processing, and the key components involved, consumers can make more informed dietary choices. This knowledge is vital for navigating the complexities of modern food systems and addressing the trade-offs between convenience and health.

Chapter 3

The Role of Additives

Additives are a cornerstone of modern food processing, playing essential roles in preserving freshness, enhancing flavor, and improving the overall appeal of processed foods. While they provide convenience and affordability, their widespread use raises questions about their long-term health effects and environmental implications. This chapter explores the functions and impacts of key categories of additives: preservatives, artificial flavors and colors, sweeteners, and sodium and fat substitutes.

Preservatives: How They Extend Shelf Life and Affect Taste

Preservatives are indispensable in modern food processing, acting as safeguards against spoilage caused by bacteria, mold, and yeast. Their primary purpose is to extend the shelf life of food products, ensuring they remain safe and consumable over time. This is especially critical in today's global food supply chains, where items must often travel long distances from production facilities to retailers and consumers. By reducing spoilage, preservatives also play a key role in minimizing food waste, a significant environmental and economic concern.

The Science of Preservation

Preservatives work by inhibiting the growth of microorganisms that cause food to decay. This is achieved through various mechanisms, such as disrupting microbial cell membranes, inhibiting enzyme activity, or altering the food's pH to create an inhospitable environment for bacteria and mold. Different preservatives are used depending on the type of food and the preservation challenges it faces.

Sodium benzoate is one of the most commonly used preservatives, particularly in acidic foods such as soft drinks, fruit juices, and pickles. Its effectiveness lies in its ability to prevent the growth of yeast and mold in low pH environments. Similarly, potassium sorbate is a widely used preservative found in baked goods, cheeses, and salad dressings. It inhibits the growth of fungi and some bacteria, ensuring that products maintain their texture and flavor over time (Moss, 2013).

In cured meats like bacon, sausages, and ham, nitrates and nitrites are used not only as preservatives but also as flavor and color enhancers. These compounds prevent the growth of *Clostridium botulinum*, the bacteria responsible for botulism, a potentially fatal illness. They also give cured meats their characteristic pink color and savory taste. However, when exposed to high heat or acidic conditions, nitrites can form nitrosamines, compounds associated with an increased risk of cancer, particularly colorectal cancer (Lahou et al., 2017). As a result, the use of nitrites has become a subject of regulatory scrutiny, and manufacturers are exploring alternative preservation methods.

Preservation of Processed Foods

Preservation techniques vary depending on the type of food. For example, in canned vegetables, preservatives such as brine (a solution of salt and water) are used to prevent microbial growth while maintaining the food's texture and flavor. While effective, this can result in a noticeably saltier taste compared to fresh alternatives. Similarly, chemical preservatives in baked goods, such as calcium propionate, prevent mold growth but may contribute to a slightly altered aftertaste, particularly in sensitive palates.

Preservation in beverages often involves the use of sulfur dioxide or its derivatives, which inhibit yeast and bacterial growth in products like wine, cider, and fruit juices. These compounds are highly effective, but they can impart a subtle chemical aftertaste, particularly if used in higher concentrations. Some consumers may also experience sensitivities to sulfur compounds, resulting in allergic-like symptoms.

Frozen foods rely less on chemical preservatives, as freezing itself inhibits microbial activity. However, some frozen products incorporate preservatives to maintain quality during thawing and prevent oxidation. For instance, frozen seafood may contain sodium tripolyphosphate to retain moisture and improve texture.

Health Implications and Controversies

While preservatives are vital for food safety and convenience, their use has sparked ongoing debate regarding potential health risks. Nitrites and nitrates, for example, have been the focus of numerous studies examining their role in the formation of nitrosamines, which are carcinogenic in high concentrations. Although regulatory agencies have set limits on acceptable levels of these compounds, consumer demand for nitrate-free and "natural" alternatives has grown, prompting innovation in the food industry.

Some preservatives, such as sodium benzoate, have also raised concerns about their potential to produce benzene, a known carcinogen, under specific conditions, such as exposure to heat and light in the presence of ascorbic acid (vitamin C). While the risk is minimal and tightly regulated, it has contributed to a growing skepticism of chemical preservatives among health-conscious consumers (Stevens et al., 2013).

Impact on Taste and Consumer Perception

Preservatives, while extending shelf life, can subtly alter the sensory qualities of food. For example, canned vegetables preserved with brine may taste saltier than their fresh counterparts, and synthetic preservatives can leave a detectable aftertaste in some processed products. These changes, though minor, can influence consumer preferences and perceptions.

Manufacturers face the challenge of balancing effective preservation with flavor retention. Advances in food technology have led to the development of "clean-label" preservatives, which are derived from natural sources such as rosemary extract, vinegar, and citrus. These alternatives aim to maintain the integrity of the food's taste while addressing consumer concerns about artificial additives.

Balancing Preservation and Nutrition

Preservatives not only extend shelf life but also protect the nutritional value of foods by preventing spoilage. Without preservatives, many perishable items would degrade rapidly, leading to significant nutrient loss. For instance, fruits and vegetables that spoil due to mold or bacteria lose valuable vitamins and antioxidants. Similarly, milk and dairy products that turn sour prematurely are rendered inedible, resulting in wasted nutritional resources.

However, the inclusion of preservatives can also have unintended consequences for nutrition. Sodium-based preservatives, for example, contribute to the overall sodium content of processed foods, which can exacerbate health issues like hypertension when consumed in excess. This underscores the importance of monitoring additive levels and promoting balanced consumption.

The Future of Food Preservation

The evolution of food preservation is driven by the dual imperatives of safety and consumer demand for natural and clean-label products. Innovations such as high-pressure processing (HPP) and pulsed electric field (PEF) technology offer chemical-free methods for extending shelf life while preserving the sensory and nutritional qualities of food. These techniques use physical processes to inactivate microbes, reducing the need for synthetic additives.

As sustainability becomes a priority, the role of preservatives in reducing food waste is increasingly recognized. By extending the usability of perishable items, preservatives contribute to more efficient food systems and help combat global food insecurity.

Preservatives are essential for modern food systems, ensuring safety, extending shelf life, and reducing waste. They enable the global distribution of food products, making them accessible to diverse populations. However, their use comes with challenges, including potential health risks and subtle impacts on taste. By balancing the benefits and trade-offs of preservatives, the food industry can continue to innovate, offering safer and more sustainable solutions that align with consumer preferences and health priorities.

Artificial Flavors and Colors: Chemical Tricks to Enhance Appeal

Artificial flavors and colors are designed to make processed foods more appealing by replicating or enhancing the sensory characteristics of natural ingredients. They are integral to creating the hyper-palatable nature of many ultra-processed foods, contributing to their widespread popularity.

Artificial flavors mimic the taste of natural ingredients using synthetic compounds. For example, vanillin is a synthetic version of vanilla, widely used in baked goods, beverages, and candies. Artificial fruit flavors, such as strawberry or banana, are often complex mixtures of esters, alcohols, and other compounds that replicate the desired taste profile. These flavors provide consistency across batches, ensuring that each product tastes the same regardless of variations in raw materials (Moss, 2013).

Artificial colors enhance the visual appeal of processed foods, influencing consumer perception of freshness and quality. Synthetic dyes like Red 40, Yellow 5, and Blue 1 are commonly used in candies, beverages, and baked goods. These colors are often derived from petroleum-based compounds and are highly concentrated, allowing for vibrant hues with minimal quantities. While approved for use by regulatory agencies, artificial colors have been associated with behavioral effects in children, such as hyperactivity, leading to calls for greater transparency and natural alternatives (Stevens et al., 2013).

The use of artificial flavors and colors raises ethical and health concerns. While they allow manufacturers to produce cost-effective and visually attractive products, their synthetic origins and potential health impacts remain subjects of debate. Increasing consumer demand for natural alternatives has driven the development of plant-based colorants and flavors, such as beet juice and turmeric extract, which offer safer and more sustainable options.

Sweeteners: High-Fructose Corn Syrup, Aspartame, and Their Biochemical Effects

Sweeteners are integral to modern processed foods, playing a central role in enhancing palatability, reducing costs, and meeting consumer demands for sweet-tasting products. They come in many forms, from natural sweeteners like sugar and honey to artificial alternatives such as high-fructose corn syrup (HFCS) and aspartame. While sweeteners provide sweetness with varying caloric values, their widespread use has raised significant concerns about their health impacts and biochemical effects on the body.

High-Fructose Corn Syrup (HFCS)

High-fructose corn syrup, derived from cornstarch, is one of the most common sweeteners used in processed foods and beverages. It is produced by enzymatically converting glucose into fructose, resulting in a sweetener with varying proportions of these two sugars. HFCS is widely used in soft drinks, baked goods, and condiments due to its affordability, stability, and ability to blend seamlessly into formulations (Bray et al., 2004).

One of the main reasons HFCS gained prominence in the food industry is its low cost. Subsidies for corn production in the United States made HFCS a cheaper alternative to cane sugar, allowing manufacturers to reduce production costs while maintaining sweetness. Its liquid form also makes it easier to transport and integrate into food products compared to granulated sugar.

However, HFCS has come under scrutiny for its potential role in the global epidemic of obesity and related metabolic disorders. Unlike glucose, which is metabolized by most cells in the body, fructose is primarily processed in the liver. Excessive consumption of fructose can overwhelm the liver, leading to increased lipogenesis (fat production) and the accumulation of fat in liver cells, contributing to non-alcoholic fatty liver disease (NAFLD). Additionally, high fructose intake has been linked to insulin resistance, a precursor to type 2 diabetes, and elevated triglyceride levels, which are associated with cardiovascular diseases (Bray et al., 2004).

HFCS consumption is also implicated in appetite regulation. Unlike glucose, fructose does not stimulate the release of insulin or leptin—hormones that signal satiety—nor does it suppress ghrelin, a hormone that stimulates hunger. This biochemical distinction can lead to overconsumption, as individuals may not feel as full after consuming HFCS-sweetened foods and beverages.

Aspartame (*sugar substitute*)

Aspartame, an artificial sweetener approximately 200 times sweeter than sugar, has been widely used in diet sodas, sugar-free gums, and low-calorie desserts since its approval in the 1980s. Its high sweetness potency means only small amounts are needed to achieve the desired sweetness, making it a popular choice for reducing calorie content in foods and beverages.

Chemically, aspartame is composed of two amino acids—phenylalanine and aspartic acid—linked by a methyl ester bond. Upon consumption, it is broken down into its components: phenylalanine, aspartic acid, and methanol, all of which occur naturally in many foods. Phenylalanine and aspartic acid are amino acids used by the body to build proteins, while methanol is metabolized into formaldehyde and formic acid, which are rapidly eliminated in small amounts.

While aspartame is considered safe for the general population, it has been subject to ongoing controversy. Some studies have raised concerns about its potential neurological effects, including headaches, dizziness, and mood changes, although these claims lack consistent evidence. Additionally, its breakdown products have led to speculation about carcinogenic risks, but comprehensive reviews by regulatory bodies, including the U.S. Food and Drug Administration (FDA) and the European Food Safety Authority (EFSA), have found no significant cancer risk at typical consumption levels (Magnuson et al., 2007).

Aspartame is not suitable for individuals with phenylketonuria (PKU), a rare genetic disorder that impairs the metabolism of phenylalanine. For this population, strict avoidance of aspartame is necessary to prevent toxic buildup.

Biochemical Effects of Sweeteners

The impacts of sweeteners extend beyond their caloric content, influencing metabolic processes, appetite regulation, and gut microbiota.

1. **Metabolic Health and Energy Balance**

   Artificial sweeteners like aspartame and HFCS are associated with metabolic disruptions. While aspartame contains negligible calories, its effects on insulin sensitivity and glucose metabolism remain debated. Some studies suggest that artificial sweeteners may influence glucose tolerance indirectly by altering gut microbiota composition (Suez et al., 2014). HFCS, on the other hand, directly contributes to calorie intake and metabolic dysregulation, as its fructose component bypasses insulin-regulated pathways and promotes fat accumulation.

2. **Appetite and Satiety**

Sweeteners, particularly artificial ones, may affect appetite regulation. The intense sweetness of artificial sweeteners can desensitize taste receptors, leading to reduced satisfaction with naturally sweet foods like fruits. This desensitization may drive increased cravings for sweet-tasting products, perpetuating a cycle of overconsumption.

3. **Gut Microbiota**

Emerging research suggests that artificial sweeteners can alter the gut microbiome, the community of microorganisms that play a vital role in digestion, immunity, and metabolism. Changes in microbiota composition have been linked to increased inflammation, impaired glucose tolerance, and weight gain, although further research is needed to clarify these effects (Suez et al., 2014).

Balancing Sweetness and Health

While sweeteners offer undeniable benefits in enhancing flavor and reducing calorie content, their potential health risks underscore the importance of moderation and informed consumption. Public health efforts have focused on reducing the intake of added sugars and promoting natural alternatives like stevia and monk fruit extract, which provide sweetness without the caloric or metabolic impacts of sugar and HFCS.

The food industry has also embraced reformulation, introducing products with reduced sugar and artificial sweeteners to meet consumer demand for healthier options. However, balancing sweetness and health remains a challenge, as many low-calorie or sugar-free products still rely on artificial additives with debated health effects.

Sweeteners like HFCS and aspartame are defining features of processed foods, providing the sweetness that consumers crave while raising questions about their long-term health implications. HFCS, though cost-effective and versatile, contributes to metabolic disorders when consumed excessively. Aspartame, while low in calories, faces ongoing scrutiny over its potential neurological and metabolic effects. Understanding the biochemical impacts of these sweeteners is essential for making informed dietary choices and guiding future innovation in food production.

Sodium and Fat Substitutes: The Hidden Chemistry in "Healthier" Processed Options

As consumers become more health-conscious, food manufacturers have developed sodium and fat substitutes to create "lighter" versions of popular products. These substitutes aim to mimic the taste and texture of traditional ingredients while reducing calories, sodium, or unhealthy fats.

Sodium substitutes, such as potassium chloride, are used to lower the sodium content of processed foods while maintaining a salty flavor. These substitutes are often found in reduced-sodium soups, snacks, and frozen meals. However, potassium chloride can impart a metallic or bitter aftertaste, which manufacturers counteract with flavor enhancers. While reducing sodium is beneficial for cardiovascular health, excessive potassium intake can pose risks for individuals with kidney disease or other conditions affecting potassium metabolism (He & MacGregor, 2007).

Fat substitutes include olestra, a synthetic fat that provides the texture and mouthfeel of fat without calories, as it passes through the digestive system undigested. Olestra was initially marketed as a breakthrough for low-fat snacks, but its side effects, including gastrointestinal distress and reduced absorption of fat-soluble vitamins, limited its popularity.

Other fat substitutes, such as modified starches and protein-based ingredients, are more widely accepted and used in products like low-fat yogurt and baked goods.

While sodium and fat substitutes offer a healthier image, they also highlight the complexity of food chemistry. These ingredients must replicate the sensory and functional properties of traditional components without compromising safety or consumer satisfaction. Balancing these factors remains a challenge for the food industry.

Additives are integral to modern food systems, enhancing safety, flavor, and convenience. Preservatives ensure food remains safe and consumable for extended periods, while artificial flavors and colors increase the sensory appeal of processed products. Sweeteners and substitutes cater to consumer demands for taste and health-conscious options. However, the widespread use of these additives raises important questions about their health effects and the ethical implications of their inclusion in processed foods.

Understanding the role of additives is a powerful tool for both consumers and the food industry. For consumers, this knowledge equips them to navigate the complexities of modern food labels, recognize the purpose and impact of various ingredients, and make dietary choices that align with their health goals. It encourages mindfulness about the balance between convenience and nutritional quality, fostering a more deliberate approach to food consumption.

For the food industry, this awareness serves as a catalyst for transparency and innovation. As consumer demand shifts toward clean-label products and natural alternatives, companies have the opportunity to reformulate their offerings, embracing safer and more sustainable additives without compromising taste, shelf life, or affordability. This evolution can rebuild trust between producers and consumers, paving the way for a food system that prioritizes health and well-being.

Balancing the benefits of additives with their potential risks is critical to addressing public health concerns while maintaining the advantages of modern food processing. By striking this balance, society can move toward a more health-conscious and sustainable approach to food production—one that respects the needs of individuals, protects the environment, and supports long-term nutritional security. This shift not only benefits current generations but also sets the foundation for a healthier and more equitable food system for the future.

Chapter 4

The Art and Science of Food Engineering

Food engineering is a multidisciplinary field that combines scientific principles and technological innovations to transform raw ingredients into processed food products. It involves precise techniques and chemical transformations that enhance texture, flavor, shelf life, and nutritional content. This chapter explores key food engineering techniques, their impacts on food properties, and the rise of "functional foods," which incorporate additional health benefits through the inclusion of vitamins, minerals, and probiotics.

Techniques: Hydrogenation, Freeze-Drying, and Extrusion

Food engineering employs sophisticated techniques to modify the physical, chemical, and sensory properties of food. These processes are designed to meet consumer demands for taste, convenience, and extended shelf life while enabling efficient mass production and distribution. Among the most impactful techniques are hydrogenation, freeze-drying, and extrusion, each playing a unique role in shaping modern food systems.

Hydrogenation

Hydrogenation is a chemical process that transforms liquid unsaturated fats into solid or semi-solid saturated fats, providing stability and functionality to a wide range of food products. This process is achieved by introducing hydrogen atoms into the carbon-carbon double bonds of unsaturated fatty acids in the presence of a metal catalyst, such as nickel, under high temperature and pressure. The result is a more stable fat that resists oxidation, reducing the likelihood of rancidity and extending the product's shelf life (Willett & Stampfer, 2013).

Hydrogenation is widely used in the production of margarine, shortening, and baked goods, where the texture and stability of fats are critical. For instance, margarine owes its spreadable consistency to partially hydrogenated oils, while shortening contributes to the flaky texture of pie crusts and pastries. This process also improves the heat tolerance of fats, making them suitable for high-temperature cooking and frying.

Despite its advantages, hydrogenation has significant health implications. Partially hydrogenated oils produce trans fats, which raise low-density lipoprotein (LDL) cholesterol ("bad" cholesterol) and lower high-density lipoprotein (HDL) cholesterol ("good" cholesterol). These effects increase the risk of cardiovascular diseases, prompting regulatory actions to limit or ban trans fats in many countries. For example, the U.S. Food and Drug Administration (FDA) declared partially hydrogenated oils as no longer generally recognized as safe (GRAS) in 2015, leading to their phased removal from food products (Willett & Stampfer, 2013).

The food industry has responded to these concerns by developing alternative processes, such as interesterification, which rearranges the fatty acids in oils to achieve similar properties without producing trans fats. These innovations reflect the evolving priorities of food engineering, balancing functionality with health considerations.

Freeze-Drying

Freeze-drying, or lyophilization, is a sophisticated preservation method that removes water from food while retaining its structure, flavor, and nutritional content. This process involves three main stages: freezing the food, reducing the surrounding pressure, and allowing the frozen water in the food to sublimate directly from solid to vapor. The result is a lightweight, shelf-stable product that can be easily rehydrated while preserving its original characteristics (Ratti, 2001).

The ability to retain the integrity of food during freeze-drying makes it particularly valuable in the production of high-value products such as instant coffee, freeze-dried fruits, vegetables, and ready-to-eat meals. Freeze-dried foods are widely used in specialized applications, including astronaut food, military rations, and emergency survival kits, where durability, low weight, and high nutritional value are paramount. For example, freeze-dried fruits like strawberries and mangoes are popular as healthy snack options that maintain their natural sweetness and texture.

Freeze-drying offers several advantages over other preservation methods. Unlike heat-based drying techniques, which can degrade heat-sensitive nutrients such as vitamin C and antioxidants, freeze-drying preserves these compounds, ensuring that the nutritional quality of the food remains intact. Additionally, the process prevents shrinkage and textural changes, making freeze-dried products visually appealing and palatable.

However, freeze-drying is an energy-intensive process, requiring specialized equipment and significant time to complete. These factors contribute to the higher cost of freeze-dried products compared to traditionally dried or canned alternatives. Despite this, the demand for freeze-dried foods continues to grow, driven by consumer preferences for convenience, quality, and health-conscious options.

Extrusion

Extrusion is a versatile mechanical process used to create a wide variety of food products, including breakfast cereals, snack foods, pasta, and pet food. The process involves forcing a mixture of ingredients through a die under high pressure and temperature, shaping the food while simultaneously cooking it. Extrusion allows for the production of uniform products with consistent textures and appearances, making it a cornerstone of modern food manufacturing (Singh et al., 2007).

The extrusion process begins with raw ingredients, such as grains, starches, and proteins, which are mixed with water and other additives to form a dough. This mixture is then fed into an extruder, where it is subjected to heat and mechanical pressure. As the dough passes through the die, it expands due to the release of pressure, creating the desired shape and texture. For example, the puffed texture of cheese snacks and the crunchy consistency of breakfast cereals are achieved through extrusion.

Extrusion offers several advantages in food production. It allows for the incorporation of flavorings, colorants, and nutritional fortifications, enabling manufacturers to create customized products that cater to specific consumer preferences. Additionally, the process reduces moisture content, enhancing the shelf stability of the final product. Extrusion is also highly efficient, minimizing waste and enabling large-scale production.

However, extrusion has limitations, particularly regarding its impact on nutritional content. The high temperatures involved can degrade heat-sensitive vitamins and amino acids, potentially reducing the overall nutritional value of the product. Efforts to mitigate these effects include optimizing processing conditions and incorporating post-extrusion fortification to replenish lost nutrients (Singh et al., 2007).

Hydrogenation, freeze-drying, and extrusion exemplify the art and science of food engineering, showcasing the innovative techniques that shape modern food products. Each process offers unique benefits, from enhancing texture and flavor to preserving nutritional content and extending shelf life. However, these techniques also pose challenges, particularly in balancing functionality with health and sustainability considerations. As consumer expectations evolve, food engineering continues to innovate, finding new ways to meet the demands of convenience, quality, and nutrition.

Extrusion is a transformative mechanical process that plays a pivotal role in modern food manufacturing, enabling the creation of diverse products such as breakfast cereals, snacks, pasta, and even pet food. This versatile technique combines shaping, cooking, and texturizing in a single operation, making it one of the most efficient and widely used methods in the food industry.

How Extrusion Works

The extrusion process begins with a mixture of ingredients—such as grains, proteins, and starches—combined with water and other additives to form a dough-like consistency. This mixture is fed into an extruder, which consists of a barrel, a screw mechanism, and a die. The screw propels the dough through the barrel under high pressure and temperature, generating heat through friction and external sources. As the dough is forced through the die at the end of the extruder, it expands due to the rapid release of pressure, forming a specific shape and texture (Singh et al., 2007).

The versatility of extrusion allows manufacturers to produce a wide range of textures, from the airy crispiness of puffed snacks to the firm chewiness of pasta. Adjustments to processing variables, such as temperature, pressure, screw speed, and die design, enable precise control over the characteristics of the final product. For example, increasing the moisture content during extrusion produces a softer texture, while higher temperatures create crispier results.

Applications of Extrusion

Extrusion is widely used to create popular consumer products, including:

1. **Breakfast Cereals:** Many cereals owe their light, crunchy texture to extrusion. Ingredients like corn, rice, and oats are processed into flakes, puffs, or rings, often with added flavorings or coatings to enhance taste and appeal.

2. **Snack Foods:** Extruded snacks such as cheese puffs, pretzels, and puffed rice are known for their uniform shapes and unique textures. Flavorings and seasonings can be evenly distributed during the extrusion process to create bold, consistent tastes.

3. **Pasta and Noodles:** Extrusion gives pasta its distinct shapes, from spaghetti to macaroni. By controlling the moisture content and extrusion temperature, manufacturers produce a firm texture that holds up during cooking.

4. **Pet Food and Animal Feed:** Extrusion is also vital in the production of pet food, creating products that are nutritionally balanced, palatable, and easily digestible.

Advantages of Extrusion

Extrusion provides numerous advantages in food manufacturing, combining efficiency, versatility, and the ability to enhance product texture, flavor, and shelf life while accommodating nutritional fortification and diverse ingredient profiles. Extrusion offers several benefits that have made it indispensable in food manufacturing:

1. **Versatility:** The process accommodates a wide variety of ingredients, including grains, proteins, fats, and dietary fibers, enabling manufacturers to produce an extensive range of products.

2. **Efficiency:** Extrusion combines multiple steps—mixing, cooking, shaping, and texturizing—into a single process, reducing production time and costs.

3. **Nutritional Fortification:** Ingredients such as vitamins, minerals, and protein isolates can be incorporated into the dough, enhancing the nutritional profile of the final product.

4. **Improved Shelf Life:** Extrusion reduces the moisture content of food, inhibiting microbial growth and increasing shelf stability without the need for additional preservatives.

Challenges and Nutritional Implications

While extrusion offers numerous advantages, it also presents challenges, particularly regarding the nutritional quality of food. The high temperatures and pressures used during extrusion can degrade heat-sensitive nutrients, such as vitamin C, thiamine, and certain antioxidants. Additionally, the process may alter the structure of proteins and carbohydrates, affecting their digestibility and glycemic index (Singh et al., 2007).

To address these issues, food engineers have developed strategies to optimize the nutritional content of extruded products. For example, post-extrusion fortification—adding heat-sensitive nutrients after processing—can mitigate nutrient losses. Advances in low-temperature extrusion technologies also aim to preserve more of the original nutritional value.

Future Directions

Extrusion continues to evolve with innovations aimed at enhancing sustainability, nutritional quality, and product diversity. Emerging trends include the use of plant-based ingredients for vegetarian and vegan products, incorporation of functional components like probiotics and dietary fibers, and the development of biodegradable packaging made from extruded materials.

By refining extrusion processes and integrating cutting-edge technologies, the food industry can create products that not only meet consumer demands for convenience and taste but also contribute to healthier and more sustainable diets

How Chemical Transformations Impact Texture, Flavor, and Nutritional Content

The chemical transformations that occur during food engineering processes are pivotal in shaping the sensory and nutritional properties of food products. These transformations, while enhancing certain qualities such as texture and flavor, often come with trade-offs that may diminish the nutritional value of the food. Balancing these outcomes is a key challenge in food engineering, as manufacturers strive to meet consumer expectations for taste and quality while maintaining nutritional integrity.

Texture: The Key to Consumer Acceptance

Texture plays a critical role in determining a food product's appeal and market success. It influences how food feels in the mouth, how it is perceived during consumption, and even its packaging and storage qualities. Food engineering techniques significantly impact texture, creating the desired mouthfeel and consistency that consumers expect.

Hydrogenation is widely used to modify the texture of fats, giving products like margarine and shortening their smooth, creamy consistency. This process solidifies fats, making them easier to spread or whip while also enhancing their stability. Similarly, extrusion is instrumental in creating the light, crispy texture of puffed snacks and breakfast cereals. By manipulating pressure and temperature during the extrusion process, manufacturers can achieve a wide range of textures, from crunchy to chewy, in a single product.

Freeze-drying, another transformative process, retains the natural structure of foods while removing moisture, resulting in a crunchy texture that is highly appealing in snacks like freeze-dried fruits and vegetables. This technique is especially valued for its ability to preserve the integrity of the original product without compromising its texture.

Chemical additives such as emulsifiers and stabilizers are also used to modify texture, creating smoother, creamier, or thicker consistencies in products like sauces, ice creams, and salad dressings. However, these additives can sometimes alter the natural mouthfeel of the food, which may be off-putting to some consumers. Striking the right balance between natural texture and chemical modifications remains a priority for food engineers.

Flavor: Enhancing Sensory Appeal

Flavor is one of the most critical attributes of food, influencing not only taste but also aroma and overall enjoyment. Chemical transformations during food engineering play a significant role in developing and stabilizing flavor profiles, ensuring that products remain consistently delicious.

The Maillard reaction, a form of non-enzymatic browning that occurs during high-heat processing, is a key technique for developing rich, savory flavors in baked goods, roasted meats, and fried foods. This reaction involves the interaction between amino acids and reducing sugars, creating complex flavor compounds and appealing aromas. It is a cornerstone of many culinary and food engineering practices.

Hydrogenation contributes to flavor stability by preventing fats and oils from becoming rancid, ensuring that products retain their intended taste over time. This stability is particularly important for long-shelf-life products like crackers, cookies, and snack bars. Similarly, extrusion allows for the uniform incorporation of flavoring agents, such as spices, herbs, and sweeteners, ensuring a consistent taste throughout the product.

However, some food engineering processes can compromise flavor. For instance, freeze-drying may cause the loss of volatile flavor compounds, which are responsible for the fresh and aromatic qualities of foods like herbs, coffee, and fruits. To counteract this, manufacturers often add natural or artificial flavorings to restore or enhance the flavor profile.

The use of flavor enhancers, such as monosodium glutamate (MSG) and artificial sweeteners, further demonstrates the role of chemical transformations in flavor development. These additives amplify taste and mask any undesirable notes introduced during processing, creating a more enjoyable sensory experience for consumers.

Nutritional Content: Enhancements and Trade-Offs

The nutritional value of food is a cornerstone of consumer health and a critical factor shaped by food engineering. The processes used in food production can enhance or diminish the nutrient profile of processed products, influencing their contribution to dietary needs. Through innovative techniques such as freeze-drying, fortification, and the use of antioxidants, food engineers aim to preserve or improve the nutritional integrity of food. However, challenges like nutrient degradation during high-temperature processing and oxidation must be addressed to strike a balance between convenience, taste, and health.

Preservation Through Freeze-Drying

Freeze-drying is one of the most effective methods for preserving and even concentrating nutrients. By removing water content through sublimation, this process maintains the integrity of essential vitamins, minerals, and antioxidants. Unlike heat-based preservation techniques, freeze-drying minimizes nutrient loss, particularly for heat-sensitive compounds like vitamin C and polyphenols. As a result, freeze-dried fruits and vegetables, such as strawberries, blueberries, and spinach, retain much of their original nutritional content, making them nutrient-dense alternatives to fresh produce (Ratti, 2001).

This preservation method also enhances the shelf life of foods without requiring chemical preservatives, catering to consumer preferences for "clean-label" products. Freeze-dried products are lightweight, easy to store, and convenient, making them popular in markets ranging from health-conscious snacks to emergency food supplies. For example, freeze-dried meals are staples in astronaut and military diets, where nutrient retention and portability are paramount.

Despite its advantages, freeze-drying can result in minor losses of volatile nutrients, such as certain B vitamins, during the initial freezing phase. Advances in freeze-drying technology aim to further minimize these losses, ensuring that freeze-dried products continue to offer superior nutritional profiles.

Fortification and Enrichment

Fortification and enrichment are powerful tools in food engineering, addressing nutrient deficiencies and improving the health value of processed foods. Fortification involves adding nutrients that are not naturally present in significant amounts, while enrichment restores nutrients lost during processing.

Extrusion plays a central role in fortification. Products like breakfast cereals, energy bars, and snack foods are often enriched with vitamins (e.g., vitamin D and folic acid), minerals (e.g., iron and calcium), and other functional ingredients like omega-3 fatty acids or dietary fiber. These additions help combat common deficiencies, such as iron-deficiency anemia and neural tube defects, particularly in vulnerable populations.

Fortification has had notable public health successes. For example, the fortification of flour with folic acid has significantly reduced the prevalence of neural tube defects in newborns in countries that mandate this practice (FAO, 2021). Similarly, the addition of iodine to table salt has nearly eradicated iodine deficiency disorders in many regions worldwide.

Challenges in fortification include ensuring nutrient stability during processing and storage. Some vitamins, such as vitamin C, are prone to degradation under heat, light, or oxygen exposure. To address this, food engineers use techniques like encapsulation, where nutrients are enclosed in protective coatings that shield them from adverse conditions. Encapsulation not only preserves the bioavailability of nutrients but also allows for their controlled release during digestion.

Nutrient Degradation During High-Temperature Processing

High-temperature processes, including frying, roasting, and extrusion, are widely used to enhance flavor, texture, and shelf life. However, these methods can also degrade heat-sensitive nutrients, leading to reduced nutritional quality in the final product.

Vitamin C, a potent antioxidant, is highly susceptible to heat and oxidation, with significant losses occurring during frying or baking. Similarly, thiamine (vitamin B1), an essential nutrient involved in energy metabolism, is degraded under prolonged heat exposure. Polyphenols, known for their antioxidant properties, also suffer reductions during high-temperature cooking, diminishing the potential health benefits of processed foods.

Food engineers are actively exploring strategies to mitigate these losses. One approach involves optimizing processing conditions, such as reducing cooking times and temperatures. Another involves the use of antioxidant additives, which can stabilize sensitive nutrients during processing and storage. Additionally, post-processing fortification allows manufacturers to reintroduce lost nutrients, ensuring the final product meets nutritional standards (Singh et al., 2007).

Oxidation and the Role of Antioxidants

Oxidation is a major challenge in food engineering, particularly for products containing unsaturated fatty acids and fat-soluble vitamins like A and E. Oxidative reactions, triggered by exposure to oxygen, light, or heat, can reduce the bioavailability of nutrients and compromise food quality.

To counteract oxidation, food engineers incorporate antioxidants into formulations. These compounds, both natural and synthetic, inhibit oxidative reactions and preserve the nutritional and sensory qualities of food. Common antioxidants include vitamin E (tocopherols), vitamin C (ascorbic acid), and synthetic compounds like butylated hydroxytoluene (BHT).

Natural antioxidants derived from plant sources, such as rosemary extract and green tea polyphenols, are gaining popularity due to consumer demand for clean-label products. These natural alternatives offer effective oxidation protection while aligning with health-conscious trends.

In addition to preserving nutrients, antioxidants play a crucial role in maintaining the sensory appeal of processed foods. For example, they prevent the rancidity of oils in snacks and baked goods, ensuring the flavor remains fresh over extended storage periods.

The Future of Nutritional Optimization

Advances in food engineering continue to focus on preserving and enhancing the nutritional content of processed foods. Emerging technologies, such as vacuum frying and low-temperature extrusion, aim to minimize nutrient losses while maintaining product quality. Encapsulation techniques are also evolving, enabling the incorporation of probiotics, omega-3 fatty acids, and other sensitive nutrients into a broader range of foods.

As consumer awareness of nutrition grows, the demand for fortified and nutrient-dense products is expected to rise. The food industry's ability to balance taste, convenience, and health will be critical in shaping the future of food systems and addressing global nutritional challenges.

Balancing Sensory and Nutritional Qualities

The chemical transformations central to food engineering enable the creation of products that cater to consumer preferences for texture, flavor, and convenience. However, these processes often come with nutritional trade-offs, as the techniques used to enhance sensory appeal may degrade heat-sensitive nutrients or alter the bioavailability of essential compounds. The challenge for food engineers is to achieve a balance, ensuring products are both delicious and healthy while addressing environmental and consumer health concerns.

Innovations in Food Engineering

Advances in food technology are redefining how sensory qualities and nutritional content are balanced. Emerging techniques are being developed to minimize nutrient loss while optimizing flavor, texture, and appearance:

1. **Low-Temperature Processing:** Unlike traditional high-heat methods, low-temperature processing reduces nutrient degradation while maintaining product quality. Techniques such as sous-vide cooking and cold-pressed juicing preserve vitamins and antioxidants, ensuring the final product is both flavorful and nutrient-dense. These methods also maintain the natural textures and colors of ingredients, appealing to health-conscious consumers.

2. **Vacuum Frying:** This innovative frying method involves cooking food under reduced pressure, which lowers the boiling point of water and allows frying at significantly lower temperatures. Vacuum frying reduces oil absorption and minimizes the formation of harmful compounds like acrylamide while preserving the natural flavor and nutritional content of foods. It is particularly popular for producing healthier snack options, such as vegetable chips and fruit crisps.

3. **Encapsulation:** Encapsulation involves coating sensitive nutrients or bioactive compounds with protective materials to shield them from adverse processing conditions, such as heat, light, or oxygen. This technique ensures the stability and

bioavailability of ingredients like probiotics, omega-3 fatty acids, and vitamins. Encapsulation also enables the controlled release of these compounds during digestion, enhancing their effectiveness and appeal in functional foods.

These innovations demonstrate the potential of food engineering to align sensory and nutritional priorities, delivering products that meet consumer expectations for taste, health, and sustainability.

The Rise of "Functional Foods"

Functional foods have emerged as a transformative trend in food engineering, catering to a growing consumer demand for products that offer health benefits beyond basic nutrition. These foods are fortified or enriched with bioactive compounds, such as vitamins, minerals, probiotics, antioxidants, and dietary fiber, to address specific health concerns and promote overall well-being.

Fortification and Enrichment

Fortification with vitamins and minerals is a cornerstone of functional food development, targeting nutrient deficiencies in populations worldwide. Products such as milk fortified with calcium and vitamin D, breakfast cereals enriched with iron and B vitamins, and orange juice with added calcium are staples of functional food markets. These interventions have played a significant role in public health:

- **Folic Acid in Bread:** The fortification of bread and cereals with folic acid has substantially reduced the incidence of neural tube defects in newborns in countries that mandate this practice. It highlights how targeted nutrient fortification can address widespread health issues effectively (FAO, 2021).

- **Iodized Salt:** Adding iodine to table salt has nearly eradicated iodine deficiency disorders, including goiter and developmental delays, in many parts of the world.

The success of these programs underscores the critical role of functional foods in improving population health. However, food engineers must ensure that fortification does not compromise sensory qualities or lead to overconsumption of certain nutrients.

Probiotics and Gut Health

Probiotics are live microorganisms that, when consumed in adequate amounts, provide numerous health benefits, particularly for digestive and immune health. Commonly incorporated into yogurts, fermented drinks, and dietary supplements, probiotics are supported by a growing body of research linking them to improved gut microbiota, reduced inflammation, and enhanced immune responses (Suez et al., 2014).

Food engineering processes ensure that probiotics remain viable during manufacturing, storage, and consumption. Techniques such as microencapsulation protect these microorganisms from heat, acidity, and other stressors, maintaining their efficacy. Probiotic-enriched products are increasingly popular among consumers seeking natural and functional solutions for digestive and overall health.

Plant-Based Functional Ingredients

Plant-based ingredients are a focal point of functional food innovation, reflecting a broader shift toward sustainability and health-conscious diets. These ingredients include phytonutrients, antioxidants, and dietary fiber, which offer a range of health benefits:

- **Phytonutrients and Antioxidants:** Found in fruits, vegetables, and herbs, these compounds combat oxidative stress and inflammation. Functional products like teas, juices, and energy bars enriched with plant-derived antioxidants cater to consumers seeking natural ways to boost health.

- **Dietary Fiber:** Fiber-enhanced products, such as cereals, snack bars, and plant-based dairy alternatives, support digestive health and help regulate blood sugar and cholesterol levels.

- **Protein-Rich Alternatives:** Plant-based protein sources, such as soy, peas, and chickpeas, are increasingly used to create functional meat substitutes and protein-enhanced snacks. These products appeal to vegetarians, vegans, and flexitarians, offering health benefits while addressing ethical and environmental concerns.

Challenges and Future Directions

While functional foods offer significant benefits, they also present challenges. Ensuring the stability and bioavailability of added nutrients is a complex task, requiring advanced techniques like encapsulation and nanotechnology. Clear labeling and evidence-based health claims are essential to building consumer trust and avoiding misleading marketing.

The future of functional foods lies in personalization. Advances in nutrigenomics and data-driven health insights are paving the way for tailored functional products designed to meet individual nutritional needs and health goals. This shift toward personalization aligns with the broader trends of precision nutrition and sustainable food systems.

Balancing sensory and nutritional qualities remains a central focus of food engineering, with innovations like low-temperature processing, vacuum frying, and encapsulation redefining the possibilities for healthier, more appealing food products. The rise of functional foods underscores the potential of food engineering to address global health challenges, offering products that go beyond basic nutrition to promote wellness and sustainability. As technology continues to evolve, the food industry is poised to meet the growing demand for products that are as healthful as they are enjoyable, shaping the future of food systems and public health.

The development of functional foods is not without challenges. Ensuring the stability and bioavailability of added nutrients during processing and storage is a complex task. Additionally, clear and accurate labeling is essential to help consumers understand the benefits of functional ingredients and avoid misleading claims.

Food engineering is both an art and a science, employing advanced techniques to transform raw ingredients into convenient, flavorful, and nutritious products. Processes like hydrogenation, freeze-drying, and extrusion have revolutionized food production, offering a wide range of textures, flavors, and preservation methods. While these techniques enhance many qualities of food, they also present challenges in maintaining nutritional integrity.

The rise of functional foods represents a promising direction for food engineering, addressing specific health needs through fortification and the inclusion of bioactive compounds. However, balancing innovation with transparency and nutritional quality remains crucial for building consumer trust and promoting long-term health.

Chapter 5

The Metabolic Fallout

The modern diet, dominated by processed foods, has profound implications for metabolic health. Sugars, refined carbohydrates, and unhealthy fats, all hallmarks of many processed products, are significant contributors to the rising prevalence of obesity, diabetes, and related disorders. These metabolic conditions often manifest through biochemical pathways associated with metabolic syndrome—a cluster of risk factors that increase the likelihood of cardiovascular disease and type 2 diabetes. Additionally, processed foods can alter the delicate balance of the gut microbiota, further exacerbating metabolic dysfunction. This chapter explores these interconnected issues in detail.

How Sugars, Refined Carbohydrates, and Unhealthy Fats Contribute to Obesity and Diabetes

The widespread consumption of processed foods, characterized by high levels of added sugars, refined carbohydrates, and unhealthy fats, has been a driving force behind the global epidemics of obesity and type 2 diabetes. These dietary components disrupt metabolic regulation through a combination of rapid energy absorption, inflammatory responses, and hormonal imbalances. Understanding their impact offers critical insights into the mechanisms linking diet to chronic disease.

Sugars and Refined Carbohydrates

Added sugars and refined carbohydrates are primary contributors to the metabolic dysregulation observed in obesity and diabetes. These ingredients are prevalent in processed foods such as sodas, candies, baked goods, and breakfast cereals.

Rapid Absorption and Blood Sugar Spikes

Added sugars like high-fructose corn syrup (HFCS) and sucrose are quickly absorbed into the bloodstream, causing rapid spikes in blood glucose levels. This triggers a corresponding surge in insulin secretion, a hormone responsible for facilitating the uptake of glucose by cells for energy or storage. Over time, the repeated stimulation of insulin pathways can lead to **insulin resistance**, where cells become less responsive to insulin, requiring higher levels of the hormone to achieve the same effect. Insulin resistance is a hallmark of type 2 diabetes and is strongly linked to the overconsumption of sugary and refined carbohydrate-rich foods (Bray et al., 2004).

Fiber Depletion and Digestive Efficiency

Refined carbohydrates, such as those found in white bread, pastries, and sugary cereals, are stripped of dietary fiber during processing. Fiber slows digestion and moderates glucose absorption, preventing sharp spikes in blood sugar. Without fiber, these carbohydrates are digested rapidly, resulting in glycemic volatility and increased demand for insulin. This cycle contributes to energy imbalances and sets the stage for chronic metabolic issues.

Caloric Surplus and Fat Storage

Excessive consumption of sugars and refined carbohydrates often leads to a caloric surplus, as these foods are calorie-dense but nutrient-poor. Surplus calories are stored as fat, particularly in visceral adipose tissue. Visceral fat, located around internal organs, is metabolically active and releases pro-inflammatory cytokines, such as tumor necrosis factor-alpha (TNF-α) and interleukin-6 (IL-6). These cytokines interfere with insulin signaling pathways, exacerbating insulin resistance and increasing the risk of cardiovascular disease (Lassale et al., 2018).

Fructose and Lipogenesis

Fructose, a component of HFCS, is metabolized primarily in the liver, bypassing the regulatory effects of insulin. When consumed in large quantities, fructose is converted into triglycerides through a process called de novo lipogenesis, contributing to the accumulation of fat in the liver (non-alcoholic fatty liver disease) and in the bloodstream. This fat accumulation further impairs insulin sensitivity and increases the risk of metabolic syndrome.

Unhealthy Fats

Unhealthy fats, including trans fats and saturated fats, compound the metabolic damage caused by sugars and refined carbohydrates. These fats are prevalent in processed foods such as margarine, fried snacks, baked goods, and processed meats.

Trans Fats and Lipid Imbalances

Trans fats are artificially created through hydrogenation, a process that solidifies liquid unsaturated fats by adding hydrogen atoms. These fats are valued in food manufacturing for their stability and long shelf life but are notorious for their adverse health effects. Trans fats raise levels of low-density lipoprotein (LDL) cholesterol, known as "bad" cholesterol, while simultaneously lowering high-density lipoprotein (HDL) cholesterol, the "good" cholesterol. This imbalance promotes the formation of arterial plaque, increasing the risk of atherosclerosis, heart attack, and stroke (Willett & Stampfer, 2013).

In addition to their effects on cholesterol, trans fats are pro-inflammatory, exacerbating systemic inflammation that interferes with insulin signaling. This inflammatory response is a key contributor to metabolic syndrome, a cluster of conditions that includes obesity, hypertension, and impaired glucose tolerance.

Saturated Fats and Insulin Sensitivity

Saturated fats, commonly found in processed meats, snack foods, and certain dairy products, also play a role in metabolic dysfunction. While moderate consumption of naturally occurring saturated fats may not pose significant risks, excessive intake from processed foods has been linked to increased lipid accumulation in the liver and skeletal muscle tissues. This accumulation interferes with insulin signaling pathways, reducing the body's ability to regulate blood sugar effectively.

Dietary Fat and Energy Density

Both trans and saturated fats are highly energy-dense, providing nine calories per gram compared to four calories per gram from carbohydrates or protein. This high energy density contributes to excessive caloric intake, leading to weight gain and the development of obesity. Obesity, in turn, amplifies the metabolic effects of unhealthy fats by increasing adipose tissue-driven inflammation and impairing insulin sensitivity.

Interconnected Impact of Sugars, Refined Carbs, and Fats

The combined consumption of sugars, refined carbohydrates, and unhealthy fats creates a synergistic effect that accelerates the progression of metabolic disorders. For example:

- Foods high in both sugars and trans fats, such as frosted donuts or cream-filled pastries, deliver a double blow to metabolic health by spiking blood sugar and promoting inflammation.

- Refined carbohydrate-based snacks fried in trans fats, such as chips, exacerbate insulin resistance through both rapid glucose absorption and lipid-driven inflammation.

These interactions illustrate the complexity of dietary contributions to obesity and diabetes, underscoring the importance of dietary patterns that limit these harmful components.

Sugars, refined carbohydrates, and unhealthy fats are pervasive in processed foods, contributing significantly to the global burden of metabolic disorders. Their effects, ranging from insulin resistance to systemic inflammation, highlight the intricate biochemical pathways through which diet influences health. Public health strategies emphasizing whole, nutrient-dense foods and reduced consumption of processed products are critical to mitigating the risks of obesity, type 2 diabetes, and related chronic conditions.

The Biochemical Pathways of Metabolic Syndrome

Metabolic syndrome is a constellation of interconnected conditions that collectively increase the risk of developing type 2 diabetes and cardiovascular disease. The syndrome includes central obesity, hypertension, dyslipidemia (abnormal cholesterol levels), and hyperglycemia. These conditions reflect underlying metabolic dysfunctions that stem from hormonal imbalances, inflammation, and oxidative stress, often exacerbated by poor dietary patterns.

Insulin Resistance and Hyperglycemia

Insulin resistance is the cornerstone of metabolic syndrome and occurs when cells in key tissues such as the liver, muscles, and adipose tissue become less responsive to insulin. Insulin, a hormone produced by the pancreas, is critical for regulating blood glucose levels by facilitating glucose uptake into cells for energy or storage.

When cells fail to respond effectively to insulin, the pancreas compensates by producing more of the hormone to maintain normal blood sugar levels. This state of hyperinsulinemia places excessive strain on pancreatic beta cells, which can eventually lead to their dysfunction. Over time, this compensatory mechanism becomes insufficient, resulting in hyperglycemia, a persistent elevation of blood glucose levels that characterize prediabetes and type 2 diabetes.

Hyperglycemia further exacerbates metabolic dysfunction by promoting the formation of advanced glycation end-products (AGEs), which damage proteins, lipids, and DNA. These AGEs contribute to oxidative stress and inflammation, worsening insulin resistance and accelerating the progression of type 2 diabetes (DeFronzo & Ferrannini, 2020).

Dyslipidemia and Hypertension

Dyslipidemia, another hallmark of metabolic syndrome, is characterized by high levels of triglycerides, low levels of HDL cholesterol, and increased small dense LDL particles. This lipid imbalance is closely tied to insulin resistance, which enhances the liver's production of very-low-density lipoproteins (VLDL) and reduces the clearance of triglycerides from the bloodstream.

Excess triglycerides are deposited in the liver and other tissues, contributing to non-alcoholic fatty liver disease (NAFLD) and systemic inflammation. NAFLD impairs liver function, further disrupting lipid metabolism and exacerbating insulin resistance. The low levels of HDL cholesterol, often referred to as "good cholesterol," hinder the removal of excess cholesterol from arterial walls, increasing the risk of atherosclerosis.

Hypertension is a common co-condition in metabolic syndrome, driven by multiple factors such as vascular inflammation, oxidative stress, and hormonal imbalances. Inflammatory cytokines and free radicals produced in the context of insulin resistance impair endothelial function, reducing nitric oxide availability. This leads to vasoconstriction and increased vascular resistance, elevating blood pressure. Diets high in sodium, often found in processed foods, compound these effects by promoting fluid retention and further increasing blood volume and pressure (Lassale et al., 2018).

Inflammation and Oxidative Stress

Chronic low-grade inflammation is a hallmark of metabolic syndrome, originating primarily from adipose tissue in individuals with central obesity. Visceral fat is metabolically active, releasing a variety of pro-inflammatory cytokines, including tumor necrosis factor-alpha (TNF-α) and interleukin-6 (IL-6). These cytokines interfere with insulin signaling pathways, creating a vicious cycle that worsens insulin resistance.

Inflammation also contributes to endothelial dysfunction, increasing the risk of atherosclerosis and cardiovascular disease. Simultaneously, oxidative stress—characterized by an imbalance between reactive oxygen species (ROS) production and antioxidant defenses—amplifies these inflammatory processes. Oxidative stress damages cellular structures, including lipids, proteins, and DNA, further impairing metabolic regulation.

The consumption of processed foods high in sugars, unhealthy fats, and artificial additives exacerbate both inflammation and oxidative stress. For example, sugary beverages and fried snacks increase the production of ROS, while trans fats contribute to systemic inflammation by activating immune cells. Together, these dietary components fuel the progression of metabolic syndrome and its complications.

Processed Foods and Their Effects on Gut Microbiota

The gut microbiota, a complex ecosystem of trillions of microorganisms, plays a crucial role in maintaining metabolic homeostasis. This diverse community aids in digestion, synthesizes essential nutrients, modulates immune responses, and communicates with distant organs through metabolic signaling. However, diets high in processed foods can disrupt this delicate balance, leading to gut dysbiosis and metabolic dysfunction.

Reduced Microbial Diversity

Processed foods are often low in dietary fiber, a critical nutrient for gut health. Fiber serves as a prebiotic, feeding beneficial gut bacteria that produce short-chain fatty acids (SCFAs) like butyrate, acetate, and propionate. SCFAs are vital for maintaining intestinal barrier integrity, reducing inflammation, and regulating glucose metabolism. A low-fiber diet diminishes SCFA production and promotes the growth of harmful bacteria, leading to gut dysbiosis (Cani et al., 2008).

Increased Intestinal Permeability

Dysbiosis can compromise the intestinal barrier, a thin layer of cells that prevents harmful substances from entering the bloodstream. This condition, commonly referred to as "leaky gut," allows bacterial endotoxins such as lipopolysaccharides (LPS) to cross into circulation. Elevated LPS levels trigger systemic inflammation, insulin resistance, and metabolic dysfunction.

Artificial Additives and Microbial Alterations

Artificial sweeteners, emulsifiers, and preservatives commonly found in processed foods also disrupt the gut microbiome. For instance, studies have shown that artificial sweeteners like saccharin and sucralose can alter gut microbial composition, impairing glucose tolerance and promoting weight gain (Suez et al., 2014). Similarly, emulsifiers have been linked to increased intestinal permeability and inflammation in animal studies.

Postbiotic and Probiotic Interventions

Addressing gut dysbiosis through dietary modifications and functional foods enriched with probiotics or prebiotics can help restore microbial balance. Probiotic strains such as *Lactobacillus* and *Bifidobacterium* have been shown to improve gut health, enhance insulin sensitivity, and reduce markers of inflammation.

Metabolic syndrome is a complex interplay of biochemical pathways influenced by diet, inflammation, and gut health. Insulin resistance, dyslipidemia, and hypertension create a cascade of metabolic disruptions, amplified by chronic inflammation and oxidative stress. The role of the gut microbiota further highlights the intricate connections between dietary patterns and systemic health. By addressing these pathways through dietary and lifestyle interventions, individuals can mitigate the risks associated with metabolic syndrome and improve long-term health outcomes.

Imbalance of Gut Microbial Diversity

The gut microbiota, a diverse community of microorganisms residing in the gastrointestinal tract, plays a central role in maintaining overall health. A balanced microbiome supports digestion, immune function, and metabolic regulation. However, diets high in processed foods are typically low in dietary fiber, a crucial nutrient that serves as a prebiotic—fuel for beneficial gut bacteria. Fiber is fermented by gut microbes to produce short-chain fatty acids (SCFAs) such as butyrate, acetate, and propionate. These SCFAs perform critical functions, including regulating inflammation, reinforcing intestinal barrier integrity, and modulating glucose and lipid metabolism (Suez et al., 2014).

Low-fiber diets, characteristic of processed food consumption, disrupt this delicate balance. Without adequate prebiotics, beneficial bacteria such as *Bifidobacterium* and *Lactobacillus* decline, while potentially harmful bacteria flourish. This shift in microbial composition, known as gut dysbiosis, reduces microbial diversity and weakens the gut's protective functions. Dysbiosis has been linked to various health issues, including inflammatory bowel disease, obesity, type 2 diabetes, and cardiovascular disease.

Diminished microbial diversity also compromises the production of SCFAs, particularly butyrate, which is essential for maintaining the health of colonocytes (intestinal epithelial cells). A lack of butyrate weakens the gut lining, making it more susceptible to damage and increasing the risk of systemic inflammation. This imbalance underscores the critical role of dietary fiber in sustaining gut health and preventing chronic diseases associated with processed food consumption.

Endotoxemia and Metabolic Inflammation

Gut dysbiosis caused by processed foods often leads to increased intestinal permeability, a condition commonly referred to as "leaky gut." The intestinal barrier is a single layer of epithelial cells held together by tight junctions, which act as a selective filter, allowing nutrients to pass into the bloodstream while blocking harmful substances. When this barrier is compromised, lipopolysaccharides (LPS)—toxins produced by certain gram-negative bacteria—can cross into systemic circulation (Cani et al., 2008).

Elevated levels of circulating LPS, a condition known as metabolic endotoxemia, trigger a cascade of inflammatory responses. LPS binds to toll-like receptor 4 (TLR4) on immune cells, activating pro-inflammatory cytokines such as tumor necrosis factor-alpha (TNF-α) and interleukin-6 (IL-6). These cytokines contribute to systemic inflammation, a key driver of insulin resistance, a hallmark of type 2 diabetes.

Chronic inflammation induced by endotoxemia also disrupts lipid metabolism, promoting the accumulation of triglycerides in the liver and adipose tissue. This exacerbates the risk of non-alcoholic fatty liver disease (NAFLD) and obesity. Furthermore, systemic inflammation impacts cardiovascular health by damaging endothelial cells, increasing the likelihood of hypertension and atherosclerosis.

The link between LPS and metabolic disorders highlights the systemic consequences of gut microbiota imbalances caused by processed food consumption. Addressing these imbalances through dietary interventions can help restore gut integrity and mitigate metabolic inflammation.

Artificial Sweeteners and Microbial Alterations

Artificial sweeteners, such as aspartame, saccharin, and sucralose, are commonly used in processed foods marketed as "low calorie" or "sugar-free." While these sweeteners offer a way to reduce calorie intake, emerging research suggests they may have unintended consequences for gut health. These compounds are not absorbed in the small intestine and pass into the colon, where they interact with gut microbes.

Studies have demonstrated that artificial sweeteners can alter the composition and diversity of the gut microbiota, promoting dysbiosis. For example, saccharin has been shown to increase the abundance of bacteria associated with inflammation, such as *Enterobacteriaceae*, while reducing beneficial strains like *Bifidobacterium* (Suez et al., 2014). These microbial shifts can impair glucose metabolism and lead to glucose intolerance, counteracting the intended benefits of using sugar substitutes.

Additionally, artificial sweeteners may increase the expression of genes involved in LPS biosynthesis, exacerbating metabolic endotoxemia. Sucralose, for instance, has been linked to reduced gut microbiota diversity and increased intestinal permeability, further contributing to systemic inflammation and insulin resistance.

Despite being classified as safe by regulatory agencies, the potential long-term effects of artificial sweeteners on gut health and metabolic function remain a topic of active investigation. These findings emphasize the need for caution in their use and highlight the importance of considering the gut microbiome in dietary choices.

The imbalance of gut microbial diversity, increased intestinal permeability, and microbial disruptions caused by artificial sweeteners illustrate the profound impact of processed food consumption on gut health. Low-fiber diets deprive beneficial bacteria of essential nutrients, leading to dysbiosis and reduced production of SCFAs, which are critical for maintaining intestinal and systemic health. The resulting endotoxemia and inflammation further contribute to metabolic disorders such as obesity, diabetes, and cardiovascular disease.

As evidence of the gut microbiome's importance in health continues to grow, strategies to promote a balanced microbiota, such as increasing dietary fiber intake and reducing reliance on artificial additives, are essential for preventing and managing chronic diseases associated with processed foods.

The metabolic fallout of processed food consumption underscores the profound impact of dietary patterns on human health. Sugars, refined carbohydrates, and unhealthy fats contribute to obesity and diabetes through biochemical pathways that disrupt metabolic regulation. Additionally, the effects of processed foods on gut microbiota further exacerbate these issues, highlighting the need for dietary strategies that prioritize whole, unprocessed foods. By understanding the metabolic pathways and mechanisms involved, public health initiatives and food engineering innovations can help mitigate the health risks associated with processed foods.

Chapter 6

The Silent Killer: Sodium

S odium is a ubiquitous ingredient in modern diets, serving crucial roles in food preservation, flavor enhancement, and processing. While essential for physiological functions, excessive sodium consumption is a major public health concern, contributing to high blood pressure, cardiovascular disease, and kidney dysfunction. The prevalence of sodium in processed foods makes it challenging for individuals to adhere to recommended intake levels, highlighting the need for systemic changes in food production and consumption practices.

Sodium's Role in Food Preservation and Flavor

Sodium, predominantly in the form of sodium chloride (table salt), has been an essential ingredient in food preservation and flavor enhancement for centuries. Its versatile properties make it indispensable in modern food production, ensuring safety, extending shelf life, and creating the savory, appealing flavors that consumers desire. However, this widespread use has also led to unintended consequences, including overconsumption and associated health risks.

Sodium as a Preservative

Sodium's role as a preservative is rooted in its ability to inhibit microbial growth by reducing water activity through osmosis. By drawing water out of food and microbial cells, sodium creates an environment that is inhospitable to spoilage organisms like bacteria, yeast, and mold. This mechanism has made sodium a critical component in preserving perishable foods and preventing foodborne illnesses.

Traditional Preservation Techniques

Historically, sodium has been central to methods like salting and brining, used to preserve meats, fish, and vegetables. For instance, cured meats such as bacon, ham, and salami rely on sodium for both preservation and flavor development. Salted fish, a staple in many cultures, demonstrates sodium's ability to extend shelf life in the absence of refrigeration.

Modern Applications

In contemporary food production, sodium-based compounds play a vital role in enhancing food safety. Sodium nitrate and sodium nitrite, for example, are commonly used in processed meats to prevent the growth of *Clostridium botulinum*, the bacteria responsible for botulism. These compounds also contribute to the characteristic pink color and flavor of cured meats. Similarly, sodium benzoate is widely used in acidic foods like salad dressings, fruit juices, and carbonated beverages to inhibit the growth of yeast and mold.

Sodium as a Flavor Enhancer

Sodium plays an indispensable role in the food industry as a powerful flavor enhancer. Beyond its technical functions in preservation and processing, sodium is a cornerstone of taste, contributing significantly to the palatability and sensory appeal of food products. Its ability to modify and amplify flavors has made it an essential ingredient across a vast range of culinary and industrial applications.

Balancing Flavors

Sodium enhances the palatability of food by balancing sweetness, masking bitterness, and intensifying umami, the savory taste associated with ingredients like meat, mushrooms, and cheese. This versatility makes it a key ingredient in processed foods, where achieving consistent and appealing flavors is essential.

Sodium-Based Flavor Enhancers

Several sodium compounds are used specifically to improve taste and texture:

- **Monosodium glutamate (MSG):** Widely recognized for its ability to impart umami, MSG is a common additive in soups, snacks, and frozen meals. It enhances savory flavors, making food more satisfying and appealing (He & MacGregor, 2010).

- **Sodium bicarbonate (baking soda):** Used in baked goods, sodium bicarbonate acts as a leavening agent, contributing to texture while subtly influencing flavor.

- **Sodium citrate:** Found in processed cheese and beverages, sodium citrate adds a tangy taste while stabilizing pH levels.

The Pervasive Use of Sodium in Processed Foods

While sodium's functional roles are invaluable, its widespread use in processed foods has led to overconsumption. Processed products, including bread, deli meats, canned soups, condiments, and snacks, are the primary sources of sodium in modern diets. Unlike the visible salt added during cooking or at the table, much of this sodium is "hidden," meaning consumers may not be aware of its presence or quantity.

Sources of Hidden Sodium:

- **Bread and Baked Goods:** Often overlooked, bread is a significant contributor to sodium intake due to the large quantities consumed regularly.

- **Deli Meats and Processed Meats:** Products like ham, bacon, sausages, and turkey slices are high in sodium for both preservation and flavor.

- **Canned and Packaged Foods:** Soups, sauces, and ready-to-eat meals rely on sodium for shelf stability and taste enhancement.

Health Implications

Most individuals in developed countries exceed the World Health Organization's (WHO) recommended daily sodium intake of 2,000 milligrams (equivalent to 5 grams of salt) without realizing it. High-sodium diets have been linked to an increased risk of hypertension, cardiovascular disease, and kidney dysfunction. The prevalence of hidden sodium in processed foods underscores the importance of raising consumer awareness and implementing strategies to reduce sodium levels in the food supply.

Sodium's dual roles as a preservative and flavor enhancer have made it indispensable in food production, from extending shelf life to creating delicious, palatable products. However, its pervasive use, particularly in processed foods, has contributed to widespread overconsumption, posing significant health risks. Addressing these challenges requires a collaborative effort between food manufacturers, policymakers, and consumers to strike a balance between sodium's functional benefits and its impact on public health.

Sodium's Impact on Blood Pressure, Cardiovascular Health, and Kidney Function

Excessive sodium consumption has significant implications for human health, particularly in the context of blood pressure regulation, cardiovascular diseases, and kidney function. While sodium is essential for maintaining fluid balance and cellular function, its overconsumption disrupts the delicate equilibrium of electrolytes in the body, leading to chronic health issues.

Blood Pressure and Hypertension

Sodium plays a critical role in regulating fluid balance by controlling the movement of water between cells and the bloodstream. This regulation is mediated by osmosis, where water flows across cell membranes to equalize sodium concentrations. When sodium intake exceeds the body's needs, water is retained to dilute the excess, increasing blood volume. This elevated blood volume places additional pressure on the walls of blood vessels, leading to hypertension (high blood pressure).

Hypertension is often referred to as the "silent killer" because it typically presents no symptoms until significant damage has occurred. Prolonged high blood pressure causes structural changes in blood vessels, including stiffening and narrowing of arterial walls. These changes increase vascular resistance, further elevating blood pressure and creating a vicious cycle. Chronic hypertension increases the workload on the heart, leading to left ventricular hypertrophy (enlargement of the heart muscle) and, ultimately, heart failure (Strazzullo et al., 2009).

High sodium intake exacerbates this cycle by impairing the kidneys' ability to excrete excess sodium, perpetuating fluid retention. The cumulative effect is a significant elevation in cardiovascular risk, particularly for heart attack and stroke.

Cardiovascular Health

Hypertension, driven by excessive sodium consumption, is a primary risk factor for cardiovascular disease. High blood pressure accelerates the development of atherosclerosis, a condition where fatty plaques build up inside arterial walls. These plaques narrow the arteries, reducing blood flow and increasing the likelihood of blood clots. The combination of narrowed arteries and elevated blood pressure significantly raises the risk of myocardial infarction (heart attack) and ischemic stroke.

Sodium also affects the function of the endothelium, the thin layer of cells lining blood vessels. The endothelium plays a crucial role in maintaining vascular health by regulating vascular tone, blood flow, and inflammation. Excessive sodium disrupts endothelial function, reducing the production of nitric oxide—a molecule that relaxes blood vessels and improves circulation. Impaired nitric oxide production leads to vasoconstriction (narrowing of blood vessels), further increasing blood pressure and contributing to cardiovascular dysfunction (He & MacGregor, 2010).

Moreover, sodium-driven inflammation and oxidative stress exacerbate vascular damage. These processes promote the instability of atherosclerotic plaques, increasing the likelihood of their rupture and subsequent clot formation. This chain of events underscores the multifaceted impact of sodium on cardiovascular health, emphasizing the need for dietary interventions to reduce sodium intake.

Kidney Function

The kidneys are vital for maintaining the body's sodium and fluid balance. When sodium intake is high, the kidneys must filter and excrete the excess to prevent fluid overload. However, chronic high sodium intake places undue stress on the kidneys, impairing their ability to function efficiently over time.

The kidneys regulate sodium levels through a complex system involving the renin-angiotensin-aldosterone system (RAAS), which adjusts blood pressure and fluid balance. Excess sodium disrupts this system, leading to sustained activation of RAAS and heightened blood pressure. This chronic strain damages the delicate nephrons, the functional units of the kidney, reducing their ability to filter waste and maintain electrolyte balance. Over time, this contributes to the development of chronic kidney disease (CKD).

Additionally, the interplay between high sodium intake and hypertension creates a feedback loop that accelerates kidney damage. Hypertension reduces blood flow to the kidneys, impairing their filtration capacity and promoting the progression of kidney disease. Sodium-related kidney dysfunction also increases the risk of end-stage renal disease (ESRD), which requires dialysis or kidney transplantation for survival (O'Donnell et al., 2014).

The Challenge of Reducing Sodium in Processed Foods

Despite widespread awareness of the health risks associated with excessive sodium consumption, reducing sodium in processed foods poses significant challenges for the food industry. Sodium's multifunctional properties make it integral to food production, and its reduction can affect flavor, texture, preservation, and safety.

Flavor and Consumer Acceptance

Sodium's role as a flavor enhancer makes it a cornerstone of processed food production. It balances sweetness, masks bitterness, and amplifies other flavors, creating the palatable profiles that consumers expect. Reducing sodium in products can result in bland or less appealing flavors, leading to decreased consumer acceptance. For example, low-sodium versions of soups, snacks, and sauces often struggle to replicate the savory depth of their traditional counterparts.

To address this challenge, food scientists are exploring alternative flavor-enhancing ingredients, such as potassium chloride and amino acids like glutamate, to mimic sodium's taste without its health risks. However, these substitutes often have limitations, including bitter aftertastes or less effective flavor enhancement. Research into microencapsulation techniques, which deliver flavor compounds in a controlled manner, shows promise in overcoming these sensory hurdles.

Preservation and Safety

Sodium's preservative properties present another obstacle to its reduction in processed foods. Sodium compounds inhibit microbial growth, ensuring food safety and extending shelf life. Reducing sodium levels can increase the risk of spoilage and contamination, particularly in perishable products like cured meats and canned goods.

Innovative preservation methods, such as high-pressure processing (HPP) and natural antimicrobial agents like rosemary extract, are being investigated as potential solutions. These techniques aim to maintain food safety while reducing reliance on sodium-based preservatives. However, implementing these methods at scale requires significant investment and research.

Technological and Economic Constraints

Reducing sodium in processed foods often involves complex reformulation processes that affect product quality and manufacturing efficiency. Sodium interacts with other ingredients to influence texture, moisture retention, and stability. For example, in baked goods, sodium contributes to dough elasticity and flavor balance. Reformulating recipes to maintain these qualities while reducing sodium levels is technically challenging and can increase production costs.

Economic constraints also play a role, as healthier low-sodium options may be more expensive to produce and less accessible to low-income populations. Collaborative efforts between public health organizations, governments, and food manufacturers are essential to address these barriers and promote equitable access to healthier food choices.

Excessive sodium consumption has profound effects on blood pressure, cardiovascular health, and kidney function, making it a significant public health concern. The challenge of reducing sodium in processed foods highlights the complexity of balancing consumer preferences, food safety, and health objectives. By advancing food science and fostering collaboration across stakeholders, it is possible to reduce sodium intake and mitigate its harmful effects, paving the way for a healthier future.

Preservation and Safety

Sodium's critical role as a preservative complicates efforts to reduce its levels in processed foods. Sodium-based compounds, particularly sodium chloride, sodium nitrate, and sodium benzoate, inhibit microbial growth by reducing water activity and creating an environment unsuitable for spoilage organisms and pathogens. For products like cured meats, cheeses, and canned goods, sodium acts as a safeguard against foodborne illnesses, such as botulism caused by *Clostridium botulinum*.

Lowering sodium levels can shorten shelf life and compromise food safety, especially in high-risk foods. For example, cured meats and processed cheeses depend on sodium not only for flavor but also for microbial inhibition. Without adequate sodium, these products are more susceptible to contamination, leading to increased spoilage and potential health risks.

To address this challenge, food scientists are exploring alternative preservation methods, including potassium chloride, which mimics sodium's preservative properties. However, potassium chloride introduces its own complications, such as a bitter or metallic aftertaste that reduces consumer acceptance. Other innovations, such as natural preservatives like rosemary extract, vinegar, and cultured dextrose, show promise but may not fully replicate sodium's multifunctional benefits. Additionally, high-pressure processing (HPP) and vacuum packaging are being employed to reduce microbial risks while minimizing sodium reliance. Despite progress, scaling these technologies to meet industrial demands requires further research and investment.

Technological and Economic Constraints

Reducing sodium in processed foods presents significant technical and economic challenges. Sodium interacts with other ingredients to influence flavor, texture, and shelf stability. Its absence can disrupt product quality, requiring complex reformulations to maintain consumer appeal.

Technical Hurdles

Sodium affects the chemical and physical properties of food. For instance, in baked goods, sodium contributes to dough elasticity and enhances the Maillard reaction, which is responsible for browning and flavor development. Replacing sodium with alternative ingredients can compromise these properties, resulting in less appealing textures, colors, and flavors. Similarly, in processed meats, sodium enhances water retention, ensuring products remain juicy and tender. Reformulating these products without sodium may lead to dryness or uneven texture.

Economic Implications

Reformulating recipes to reduce sodium often increases production costs due to the need for specialized ingredients or advanced processing technologies. For example, alternatives like potassium chloride or encapsulated flavor enhancers are more expensive than traditional sodium-based compounds. These costs are typically passed on to consumers, making low-sodium options less accessible to low-income populations who already face barriers to healthy eating.

In addition, sodium reduction initiatives require investment in consumer education to build awareness about the importance of low-sodium diets and the taste differences in reformulated products. Balancing these costs while maintaining affordability and quality is a persistent challenge for manufacturers.

Policy and Public Health Efforts

Recognizing the public health implications of high sodium consumption, governments and health organizations worldwide have implemented strategies to reduce sodium intake at the population level. These efforts focus on collaboration between regulatory agencies, food manufacturers, and consumers to create a healthier food environment.

Front-of-Package Labeling

Nutritional labels that clearly indicate sodium content help consumers make informed dietary choices. Countries like Chile and Mexico require warning labels on high-sodium products, which have effectively reduced consumption of unhealthy foods. Similar initiatives in other regions encourage food manufacturers to reformulate products to meet consumer demand for healthier options.

Sodium Reduction Targets

Many governments set voluntary or mandatory sodium reduction targets for food manufacturers. For example, the United Kingdom's salt reduction program, initiated in 2003, established clear guidelines for sodium levels across various food categories. Over a decade, this program reduced average sodium intake by 15%, resulting in measurable declines in blood pressure and cardiovascular disease mortality (He et al., 2014). Such coordinated efforts demonstrate the potential of regulatory policies to drive meaningful change.

Public Awareness Campaigns

Educational campaigns play a critical role in shifting consumer behavior. Initiatives like the American Heart Association's "Salty Six" highlight common high-sodium foods, empowering individuals to reduce their intake. Public messaging about the health risks of sodium encourages consumers to seek low-sodium alternatives, prompting manufacturers to innovate and adapt.

Sodium is both a functional necessity in food production and a significant contributor to chronic disease when consumed in excess. Its dual role highlights the complexity of addressing sodium-related health risks. While sodium enhances flavor, texture, and safety, its pervasive use in processed foods drives overconsumption, leading to hypertension, cardiovascular disease, and kidney dysfunction.

Reducing sodium intake requires a multifaceted approach that combines food industry innovation, regulatory action, and consumer education. Advances in alternative preservation methods, reformulation strategies, and public health initiatives are crucial for creating a food environment that prioritizes health without sacrificing quality. By fostering healthier dietary habits and supporting sodium reduction efforts, society can mitigate the silent but pervasive impact of sodium on public health.

Sodium plays an indispensable role in food production, serving as a preservative, flavor enhancer, and essential component of many processed foods. However, its overuse has become a silent driver of chronic diseases, including hypertension, cardiovascular disorders, and kidney dysfunction. Balancing sodium's functional benefits with its health risks requires innovation, collaboration, and awareness at every level—from food manufacturers and policymakers to consumers.

By embracing technological advancements, reformulating products, and implementing effective public health strategies, we can reduce sodium intake without compromising food quality or safety. Achieving this balance is critical for fostering healthier dietary habits and mitigating the widespread health burden associated with excessive sodium consumption. The future of sodium reduction lies in a collective effort to prioritize well-being while maintaining the functionality and palatability of our food supply.

Chapter 7

Chemical Additives and Controversies

Chemical additives are integral to modern food production, serving roles that enhance flavor, extend shelf life, and improve the overall safety and convenience of food products. However, their widespread use has sparked significant debate and concern regarding their potential health effects. This chapter delves into the controversies surrounding common additives like MSG, BPA, and artificial sweeteners, as well as emerging issues related to microplastics and chemical migration from packaging. Additionally, it examines the challenges of defining "safe levels" for additive exposure and the cumulative risks associated with long-term consumption.

Debates Over Additives: MSG, BPA, and Artificial Sweeteners

Chemical additives are pivotal in modern food production, enhancing flavor, extending shelf life, and providing alternatives for health-conscious consumers. However, the widespread use of certain additives has raised significant health concerns, leading to heated debates about their safety and long-term effects. Among the most controversial are monosodium glutamate (MSG), bisphenol A (BPA), and artificial sweeteners like aspartame and saccharin. These additives have sparked widespread scrutiny, scientific inquiry, and regulatory challenges (see table below).

Additive	Purpose	Controversy
Monosodium Glutamate (MSG)	Flavor enhancer (umami taste)	Stigmatized due to alleged health effects like 'Chinese Restaurant Syndrome'; concerns over sodium consumption.
Bisphenol A (BPA)	Used in food packaging (cans, plastics)	Endocrine disruptor; potential reproductive and developmental issues.
Artificial Sweeteners (e.g., Aspartame, Saccharin, Sucralose)	Sugar substitutes in low-calorie products	Potential gut microbiota disruption, metabolic effects, and possible neurological impact.
High-Fructose Corn Syrup (HFCS)	Sweetener in processed foods and beverages	Linked to obesity, diabetes, and metabolic syndrome.
Sodium Nitrate and Sodium Nitrite	Preservatives in cured meats	Potential carcinogen; associated with increased cancer risks.
Food Dyes (e.g., Red 40, Yellow 5, Blue 1)	Artificial coloring in food	Linked to hyperactivity in children and possible carcinogenicity.
Trans Fats (Partially Hydrogenated Oils)	Enhances shelf life and texture	Strongly linked to heart disease; banned in the U.S.
Propyl Paraben	Preservative in baked goods	Possible endocrine disruptor may affect fertility and hormone levels.
Potassium Bromate	Strengthens dough and improves bread texture	Possible human carcinogen; banned in many countries.
Carrageenan	Thickener and stabilizer in dairy products	Linked to gastrointestinal inflammation and digestive issues.
Brominated Vegetable Oil (BVO)	Emulsifier in citrus-flavored soft drinks	Contains bromine; may cause thyroid dysfunction and neurological problems.

Among the countless chemicals used in modern industry, two additives stand out as pivotal yet controversial. These substances are widely employed for their effectiveness in enhancing product performance, but their environmental and health implications are increasingly drawing scrutiny. From their persistence in ecosystems to their potential toxicity to humans, these additives epitomize the complex balance between industrial advancement and sustainable responsibility. Understanding their impacts is crucial to addressing the broader challenges posed by chemical additives in our modern world. Here are some examples:

Monosodium Glutamate (MSG)

MSG is a sodium salt of glutamic acid, a naturally occurring amino acid found in many foods, including tomatoes, cheese, and soy sauce. It is widely used as a flavor enhancer in processed foods, soups, snacks, and Asian cuisine, where it imparts the savory umami taste.

The Controversy

The debate surrounding MSG began in the 1960s with reports of "Chinese Restaurant Syndrome," a collection of symptoms including headaches, flushing, and nausea allegedly triggered by MSG consumption. While these reports gained media attention and fueled public fear, scientific studies have largely failed to establish a consistent link between MSG and these adverse effects (He et al., 2010).

Despite being classified as safe by the U.S. Food and Drug Administration (FDA) and the European Food Safety Authority (EFSA), MSG remains stigmatized, often avoided due to lingering misconceptions. Critics argue that its widespread use may contribute to overconsumption of sodium and unhealthy eating patterns, as MSG is frequently added to calorie-dense, nutrient-poor processed foods.

Current Perspective

Scientific consensus maintains that MSG is safe for consumption at typical dietary levels, with adverse effects observed only at exceptionally high doses. However, its association with highly processed foods continues to raise concerns about its indirect contribution to obesity and metabolic disorders.

Bisphenol A

Bisphenol A (BPA) is a synthetic industrial chemical that has been widely used in the production of polycarbonate plastics and epoxy resins. These materials are prevalent in many consumer goods, particularly in food and beverage packaging, where they contribute to the durability and functionality of containers. BPA is often found in the lining of metal food and beverage cans, water bottles, storage containers, and some plastic wraps. The chemical is utilized for its ability to create strong, lightweight, and heat-resistant packaging that can withstand the stresses of handling, storage, and transportation.

Polycarbonate plastics containing BPA are commonly used in products that require transparency and strength, such as reusable water bottles, food containers, and baby bottles. Epoxy resins made with BPA are used as coatings inside metal cans to prevent corrosion, preserve the quality of the food inside, and extend the shelf life of canned goods. The inclusion of BPA in these materials ensures that the packaging remains intact and functional over time, making it a popular choice for the food and beverage industry.

Despite its widespread use, BPA has raised significant health concerns due to its potential as an endocrine disruptor. Studies have suggested that BPA can leach into food and beverages from packaging, especially when containers are exposed to heat or acidic substances. This has led to growing concerns about its potential effects on human health, particularly in relation to hormone regulation, fertility, and developmental issues in children. As a result, several countries have moved toward reducing or eliminating BPA use in certain products, especially those intended for infants and young children, and have introduced regulations to limit BPA exposure in consumer goods.

The Controversy

Concerns over BPA center on its potential as an endocrine disruptor, a compound that mimics or interferes with the body's hormonal systems. Studies have linked BPA exposure to reproductive issues, developmental abnormalities, and increased risks of breast and prostate cancer. Evidence suggests that BPA can leach into food and beverages, particularly when containers are heated or damaged, raising questions about its safety (Rubin, 2011).

While regulatory agencies, including the FDA and EFSA, have concluded that BPA is safe at current exposure levels, independent studies and advocacy groups argue for stricter regulations or outright bans. Canada and the European Union, for example, have banned BPA in baby bottles, citing precautionary principles.

Emerging Evidence

Recent research continues to explore BPA's low-dose effects and its potential impact on vulnerable populations, such as infants and pregnant women. Alternatives to BPA, like bisphenol S (BPS), are increasingly used, but their safety profiles remain under scrutiny, as early studies suggest similar endocrine-disrupting properties.

Artificial Sweeteners: A Double-Edged Sword

Artificial sweeteners such as aspartame, saccharin, and sucralose have become popular sugar substitutes, offering sweetness without the caloric content of sugar. These compounds are widely used in diet sodas, sugar-free snacks, and diabetic-friendly products, marketed as tools for weight management and blood sugar control.

The Controversy

Despite their benefits, artificial sweeteners have faced criticism over their potential health risks. Concerns include:

1. **Gut Microbiota Disruption**

   Studies have shown that artificial sweeteners like saccharin and sucralose can alter the composition of gut microbiota, impairing glucose tolerance and potentially contributing to metabolic disorders (Suez et al., 2014). These findings challenge the assumption that artificial sweeteners are inert and highlight their complex interaction with human biology.

2. **Cancer Risk**

Early studies in the 1970s linked saccharin to bladder cancer in laboratory rats, leading to warnings on product labels. Subsequent research found that the mechanisms observed in rats did not apply to humans, resulting in the removal of saccharin from the National Toxicology Program's list of carcinogens. However, the specter of cancer risk continues to influence public perception.

3. **Neurological Effects**

Aspartame, metabolized into phenylalanine, aspartic acid, and methanol, has been scrutinized for its potential neurological effects. While current evidence suggests no significant risks at typical consumption levels, anecdotal reports of headaches and dizziness persist, fueling skepticism about its safety (Magnuson et al., 2007).

Do yourself a favor: flip that package of "No Sugar" or "healthy alternative" food, candy, or snack around, and Google every single ingredient. Go ahead, I'll wait. By the time you're done, you'll be so horrified by the unpronounceable science experiments lurking in your snack that you might just start gnawing on raw meat and harvesting kale from your neighbor's garden. Bon appétit!

Current Perspective

Regulatory agencies, including the FDA and WHO, endorse artificial sweeteners as safe within established acceptable daily intake (ADI) levels. However, the long-term effects of chronic exposure and their role in influencing taste preferences and dietary habits remain areas of active research.

Emerging Concerns: Microplastics and Chemical Migration from Packaging

As packaging technologies advance to meet the demands of convenience, food safety, and sustainability, new concerns have emerged regarding the unintended introduction of harmful substances into the food supply. Among the most pressing issues are the pervasive presence of **microplastics** and the migration of chemical compounds from packaging materials into food and beverages. These concerns highlight the complex interplay between innovation in food packaging and its potential health risks.

Microplastics

Microplastics, tiny plastic particles less than 5 millimeters in size, have become an alarming symbol of the global plastic pollution crisis. Once confined to environmental discussions, their presence has now infiltrated the food we eat and the water we drink, raising urgent questions about their impact on human health. Found in products ranging from seafood to bottled water, microplastics enter the food chain through a variety of pathways, including marine ecosystems, agricultural practices, and even airborne particles settling on food surfaces. Their pervasive nature underscores the hidden consequences of our reliance on plastic and highlights the need for greater awareness and action to mitigate their presence in our diets and environments.

Definition and Sources

Microplastics are small plastic particles, less than 5 millimeters in size, that originate from the breakdown of larger plastic items or as byproducts of industrial processes. Primary microplastics are intentionally manufactured for products like cosmetics and industrial abrasives, while secondary microplastics result from the degradation of larger plastic waste due to environmental exposure. They have been detected in various foods and beverages, including seafood, table salt, and bottled water, as well as in atmospheric particles that settle on open food.

Pathways into the Food Supply

Marine environments are a significant pathway for microplastics to enter the food chain. Fish, shellfish, and other marine organisms inadvertently ingest microplastics, which accumulate in their tissues. Humans consuming these organisms are at risk of secondary ingestion. Agricultural practices also contribute, as microplastics in soil from plastic mulch or biosolids used as fertilizer can enter plant systems, further contaminating the food supply.

Health Impacts

The potential health effects of microplastics are a growing area of research. Studies suggest that microplastics can accumulate in the gastrointestinal tract, causing inflammation, disrupting gut microbiota, and potentially leading to systemic exposure. Microplastics can also act as carriers for toxic chemicals, including phthalates and bisphenols, which leach from the plastics themselves. These chemicals are linked to endocrine disruption, immune suppression, and developmental issues (Barboza et al., 2018). Additionally, nanoparticles derived from plastics may penetrate cell membranes, raising concerns about their effects on human organs and cellular functions.

Mitigation Strategies

Addressing the pervasive presence of microplastics requires systemic interventions. Governments and industries are exploring regulations to reduce plastic waste, improve recycling processes, and promote the use of biodegradable alternatives. Innovations in packaging technology, such as edible films and biodegradable polymers, aim to minimize microplastic contamination. Public awareness campaigns are also essential to encourage reduced reliance on single-use plastics and advocate for environmentally responsible choices.

Chemical Migration from Packaging

Chemical migration from food packaging is an often-overlooked but critical aspect of food safety. This phenomenon involves the transfer of substances from packaging materials into the food or beverages they contain. Influenced by factors such as temperature, storage duration, and the chemical composition of the packaging and food, migration can introduce a range of compounds into the human diet. Common chemicals involved in this process include phthalates, bisphenols, and per- and polyfluoroalkyl substances (PFAS), many of which have been linked to adverse health effects. As packaging technologies evolve to meet the demands of convenience and sustainability, understanding and mitigating chemical migration has become essential for safeguarding public health.

Definition and Mechanisms

Chemical migration refers to the transfer of substances from food packaging materials into the food or beverage they contain. This migration is influenced by factors such as the chemical composition of the packaging, the nature of the food (e.g., acidity or fat content), storage conditions (e.g., temperature or duration), and mechanical stress on the packaging material.

Common Migrating Chemicals

- **Phthalates:** Widely used as plasticizers, phthalates improve the flexibility and durability of plastics. However, they are non-covalently bound to plastic structures, allowing them to leach into foods, particularly those with high-fat content. Phthalates are associated with endocrine disruption, reproductive toxicity, and potential impacts on child development.

- **Per- and Polyfluoroalkyl Substances (PFAS):** Found in grease-resistant coatings used for fast food wrappers, microwave popcorn bags, and pizza boxes, PFAS are known as "forever chemicals" due to their environmental persistence.

These substances are linked to immune system suppression, developmental delays, and an increased risk of certain cancers.

- **Bisphenol A (BPA) and Alternatives:** Common in polycarbonate plastics and epoxy resin coatings for cans, BPA can leach into food and beverages under conditions of heat, acidity, or prolonged storage. Alternatives like bisphenol S (BPS) are being introduced, but early studies suggest they may pose similar risks.

Health Impacts

Chemical migration presents a unique challenge due to its cumulative nature. Exposure to low levels of multiple chemicals over time can result in a "cocktail effect," where the combined impact is greater than the sum of individual exposures. Vulnerable populations, such as infants, pregnant women, and individuals with pre-existing health conditions, are particularly at risk. For instance, PFAS exposure has been associated with immune dysregulation and vaccine efficacy reduction, while chronic exposure to phthalates is linked to metabolic syndrome and hormone-related cancers (Vandenberg et al., 2012).

Regulatory and Industry Responses

Regulatory bodies like the FDA and EFSA set limits on permissible levels of chemical migration to ensure food safety. However, these standards often focus on individual chemicals rather than their combined effects. Industry efforts to address these concerns include:

- Developing safer packaging materials that reduce chemical migration risks.

- Increasing the use of recycled content in packaging with rigorous safety testing.

- Exploring green chemistry principles to design non-toxic packaging alternatives.

Understanding "Safe Levels" and the Cumulative Effect of Long-Term Exposure

The concept of "safe levels" for chemical additives and contaminants is foundational to regulatory frameworks, but it presents significant challenges in practice. These thresholds are determined based on toxicological studies and are designed to minimize acute and chronic risks. However, real-world exposure scenarios often deviate from controlled laboratory conditions.

Regulatory Frameworks

Regulatory agencies like the FDA, EFSA, and WHO rely on acceptable daily intake (ADI) limits, which are derived from animal studies and incorporate safety margins. While these limits aim to protect the general population, they often fail to address:

- **Cumulative Exposure:** People are exposed to multiple chemicals from various sources, including food, water, air, and consumer products. The cocktail effect complicates risk assessment, as interactions between chemicals may amplify their toxic effects.

- **Vulnerable Populations:** Infants, children, pregnant women, and individuals with compromised health may have lower thresholds for adverse effects, necessitating more stringent protections.

Bioaccumulation and Chronic Exposure

Some chemicals, such as PFAS and BPA, are lipophilic, meaning they accumulate in fatty tissues over time. This bioaccumulation increases the risk of chronic health effects, even at low exposure levels. Long-term exposure to endocrine-disrupting chemicals, for instance, has been implicated in obesity, diabetes, and hormone-related cancers. Additionally, the disruption of gut microbiota by migrating chemicals can have far-reaching effects on immune function, metabolism, and mental health.

Future Directions

To address these challenges, scientists and policymakers are advocating for:

- **Holistic Risk Assessment:** Moving beyond single-chemical analyses to evaluate the combined effects of multiple exposures.

- **Improved Consumer Transparency:** Enhancing food labeling to include information about potential chemical migration and associated risks.

- **Advancements in Packaging Technology:** Developing materials that are both functional and non-toxic, including bio-based plastics and coatings derived from natural sources.

The emerging concerns of microplastics and chemical migration from packaging materials highlight the unintended consequences of technological advancements in food production. While these innovations offer convenience and functionality, their potential health impacts demand urgent attention. By fostering interdisciplinary collaboration between scientists, regulators, and industry leaders, it is possible to mitigate these risks and ensure a safer, more sustainable food system.

Bioaccumulation and Chronic Exposure

Certain chemicals, such as BPA and PFAS, can accumulate in the body over time, increasing the risk of adverse effects even at low exposure levels. Chronic exposure to endocrine disruptors, for instance, has been linked to metabolic disorders, hormone-related cancers, and developmental abnormalities. Additionally, long-term consumption of artificial sweeteners may alter gut microbiota, potentially affecting glucose metabolism and immune function.

Public Health Implications

Understanding the cumulative impact of chemical additives requires a shift in regulatory paradigms to emphasize aggregate risk rather than isolated exposure. Public health initiatives should also prioritize consumer education, transparent labeling, and the promotion of minimally processed foods to reduce additive consumption.

The controversies surrounding chemical additives reflect the broader tension between technological innovation and public health. While additives like MSG, BPA, and artificial sweeteners have revolutionized food production, their potential risks highlight the need for rigorous oversight, transparent communication, and ongoing research. Emerging concerns, such as microplastics and chemical migration, further underscore the importance of rethinking packaging and additive use in the context of long-term health and sustainability. By addressing these challenges, society can strike a balance between the benefits of modern food technologies and the imperative to protect public health.

Chapter 8

Processed Foods and Chronic Diseases

Processed foods are a dominant feature of modern diets, providing convenience, affordability, and extended shelf life. However, their pervasive consumption has been strongly linked to the rising prevalence of chronic diseases, including cancer, heart disease, and autoimmune disorders. This chapter examines the mechanisms by which processed foods contribute to these health issues, particularly through the inflammatory responses triggered by chemical additives and refined ingredients. It also explores the broader role of processed foods in the global epidemic of non-communicable diseases (NCDs), offering insights into how dietary shifts are shaping public health outcomes worldwide.

Links to Cancer, Heart Disease, and Autoimmune Disorders

Processed foods are increasingly implicated in the development of chronic diseases, with their nutrient-poor profiles and high levels of harmful compounds contributing to the progression of several serious health conditions. The links to cancer, heart disease, and autoimmune disorders highlight the multifaceted risks posed by a diet dominated by ultra-processed products.

Cancer

Diets rich in ultra-processed foods are associated with an increased risk of certain cancers, including colorectal, breast, and gastric cancers. This connection stems from several mechanisms, primarily the presence of carcinogenic compounds that emerge during food processing and preservation.

Carcinogenic Compounds in Processing

Ultra-processed foods often contain substances that form during high-temperature cooking, such as polycyclic aromatic hydrocarbons (PAHs), heterocyclic amines (HCAs), and acrylamide:

- **PAHs and HCAs** are produced when proteins in meat are exposed to high temperatures during grilling, frying, or roasting. These compounds are mutagenic, meaning they can directly damage DNA, increasing the risk of cancer.

- **Acrylamide**, formed when starchy foods like bread, potato chips, and fried snacks are cooked at high temperatures, has been classified as a probable human carcinogen by the International Agency for Research on Cancer (IARC). Studies have shown that chronic exposure to acrylamide may elevate the risk of gastrointestinal and endocrine cancers (Friedman, 2003).

Processed Meats and Nitrosamines

Processed meats, such as bacon, sausages, and hot dogs, are preserved using nitrates and nitrites, which inhibit bacterial growth and enhance flavor. However, during cooking or digestion, these compounds can convert into nitrosamines, potent carcinogens linked to colorectal cancer. In fact, the IARC classifies processed meats as Group 1 carcinogens, placing them in the same category as tobacco and asbestos due to their strong association with cancer risk (Lahou et al., 2017).

Emerging Concerns

- **Artificial Additives and Colorings:** Some additives, such as artificial colors and preservatives, are under investigation for their potential carcinogenic effects, particularly when consumed in large quantities over long periods.

- **Endocrine-Disrupting Chemicals:** Processed foods often come in contact with packaging materials that leach bisphenol A (BPA) and phthalates, both of which have been linked to hormone-driven cancers, such as breast and prostate cancer.

Dietary Implications

The role of ultra-processed foods in cancer development underscores the importance of dietary quality. Diets rich in whole, unprocessed foods, such as fruits, vegetables, whole grains, and lean proteins, provide antioxidants, fiber, and phytochemicals that combat oxidative stress and inflammation, reducing cancer risk.

Heart Disease

Processed foods are a significant contributor to the global burden of cardiovascular disease (CVD), the leading cause of mortality worldwide. Their high content of trans fats, sodium, and refined sugars disrupts cardiovascular health by influencing cholesterol levels, blood pressure, and metabolic processes, ultimately increasing the risk of conditions like hypertension, atherosclerosis, and myocardial infarction.

Trans Fats: The Silent Threat

Trans fats, often present in margarine, fried foods, baked goods, and snack items, are created during the industrial hydrogenation of vegetable oils. This process alters the chemical structure of unsaturated fats, making them solid at room temperature and more stable for extended shelf life.

- **Impact on Cholesterol Levels**

  Trans fats raise low-density lipoprotein (LDL) cholesterol, often referred to as "bad cholesterol," while simultaneously lowering high-density lipoprotein (HDL) cholesterol, or "good cholesterol." Elevated LDL levels contribute to the buildup of fatty plaques within arterial walls, a condition known as atherosclerosis. Reduced HDL levels hinder the removal of cholesterol from arteries, further exacerbating plaque formation.

- **Promotion of Inflammation**

  Trans fats are also pro-inflammatory, stimulating the release of cytokines and other inflammatory mediators that damage blood vessels. This chronic inflammation accelerates the progression of cardiovascular disease, making trans fats one of the most harmful components of processed foods.

Recognizing these risks, many countries have implemented bans or regulations to limit trans fats in food products. However, trans fats remain a concern in regions with less stringent food safety standards, where they continue to contribute significantly to cardiovascular morbidity and mortality.

Sodium: A Catalyst for Hypertension

Sodium, commonly added to processed foods for flavor enhancement and preservation, is another major contributor to heart disease. Excessive sodium intake disrupts the body's fluid balance, leading to increased blood volume and elevated blood pressure.

- **Hypertension and Cardiovascular Risk**

  High sodium levels cause the body to retain water, which increases blood pressure, placing strain on the cardiovascular system. Prolonged hypertension damages arterial walls, promoting their stiffening and narrowing. This condition, known as arteriosclerosis, increases the risk of stroke, heart failure, and other cardiovascular events (He & MacGregor, 2010).

- **Sodium-Rich Processed Foods**

  Common processed foods like canned soups, deli meats, snack foods, and frozen meals are primary sources of dietary sodium. These foods contribute disproportionately to sodium consumption, often exceeding the recommended daily limit of 2,300 milligrams set by health organizations.

Efforts to reduce sodium intake have included public health campaigns, reformulation of food products, and front-of-package labeling. Despite these initiatives, sodium consumption remains high globally, highlighting the need for stricter regulations and consumer education.

Refined Sugars and Simple Carbohydrates: A Metabolic Double-Edged Sword

Refined sugars and simple carbohydrates, abundant in processed foods such as pastries, sugary beverages, and breakfast cereals, are detrimental to heart health due to their effects on insulin resistance, inflammation, and obesity.

- **Insulin Resistance and Obesity**

  The rapid absorption of refined sugars into the bloodstream causes spikes in blood glucose levels, triggering excessive insulin release. Over time, cells become resistant to insulin's effects, a condition known as insulin resistance, which is a precursor to type 2 diabetes and obesity. Excess body fat, particularly visceral fat, increases the risk of heart disease by releasing pro-inflammatory substances and disrupting lipid metabolism.

- **Promotion of Atherosclerosis**

  Chronic consumption of refined sugars contributes to the development of atherosclerosis, where fatty deposits accumulate in arterial walls. This process reduces blood flow to the heart and brain, significantly elevating the risk of myocardial infarction (heart attack) and stroke.

- **Inflammatory Responses**

  Refined sugars also stimulate the production of pro-inflammatory markers such as C-reactive protein (CRP) and interleukin-6 (IL-6), both of which are linked to endothelial dysfunction and vascular damage.

The Synergistic Impact of Processed Food Components

The combined effects of trans fats, sodium, and refined sugars in processed foods create a synergistic risk for heart disease. These components not only act independently to damage cardiovascular health but also amplify each other's effects when consumed together. For example:

- High sodium levels exacerbate hypertension, which interacts with trans-fat induced atherosclerosis to increase the likelihood of heart failure.

- Refined sugars promote obesity, a condition that worsens the inflammatory effects of trans fats and sodium on vascular health.

This interconnectedness underscores the importance of addressing processed food consumption, rather than focusing on individual components.

Policy and Lifestyle Interventions

Efforts to mitigate the impact of processed foods on heart disease include both policy measures and individual lifestyle changes:

- **Policy Measures**

    Governments worldwide have introduced regulations to reduce trans fats, implement sodium reduction targets, and impose taxes on sugary beverages. These measures have shown promise in lowering the prevalence of cardiovascular risk factors.

- **Lifestyle Changes**

    Shifting toward a diet rich in whole, unprocessed foods, including vegetables, fruits, whole grains, lean proteins, and healthy fats, can significantly reduce the risk of heart disease. Educating consumers about reading food labels and understanding the health implications of processed foods is equally critical.

Processed foods are a leading contributor to heart disease, driven by their high content of trans fats, sodium, and refined sugars. These components disrupt cholesterol levels, blood pressure, and metabolic health, creating a perfect storm for cardiovascular dysfunction. Addressing this public health crisis requires a multifaceted approach that combines regulatory action, public health initiatives, and consumer education to promote healthier dietary habits and reduce the global burden of cardiovascular disease.

Autoimmune Disorders

Autoimmune disorders, such as rheumatoid arthritis (RA), lupus, and inflammatory bowel disease (IBD), arise when the immune system mistakenly attacks the body's own tissues. These chronic and often debilitating conditions are influenced by genetic, environmental, and dietary factors. Emerging research suggests that the consumption of processed foods may play a significant role in both the development and progression of autoimmune disorders. The mechanisms underlying this connection often involve gut permeability, or "leaky gut," and disruptions in the balance of gut microbiota caused by food additives and other components of processed foods.

Leaky Gut and Immune Dysregulation

The intestinal lining serves as a critical barrier, preventing harmful substances such as toxins, undigested food particles, and pathogens from entering the bloodstream. In individuals with autoimmune conditions, this barrier is often compromised, leading to increased intestinal permeability, or "leaky gut."

- **Mechanism of Leaky Gut**

  When the gut barrier is weakened, tight junctions between intestinal cells loosen, allowing substances to pass through that would normally be blocked. These substances can activate the immune system, triggering an inflammatory response. Over time, chronic inflammation may lead to immune dysregulation, increasing the likelihood of autoimmune reactions (Fasano, 2012).

- **Processed Foods and Leaky Gut**

  Certain components of processed foods, such as emulsifiers and artificial sweeteners, have been shown to damage the gut lining and alter its permeability. For example:

  o **Emulsifiers**, commonly found in products like margarine, ice cream, and baked goods, disrupt the mucous layer that protects the gut lining. Studies suggest that emulsifiers like carboxymethylcellulose and polysorbate-80 can weaken the gut barrier, promoting inflammation and microbial imbalances (Chassaing et al., 2015).

  o **Artificial Sweeteners**, such as aspartame and saccharin, are associated with gut microbiota alterations that exacerbate immune responses.

Disruption of the gut microbiome can impair the regulatory functions of the immune system, increasing the risk of autoimmune conditions (Suez et al., 2014).

Gut Microbiota and Autoimmunity

The gut microbiota, a diverse community of microorganisms residing in the gastrointestinal tract, plays a central role in maintaining immune homeostasis. A healthy microbiome produces short-chain fatty acids (SCFAs) like butyrate, which help regulate inflammation and maintain gut barrier integrity. Disruptions in this delicate balance, often caused by processed foods, can contribute to autoimmune disease.

- **Microbiota Dysbiosis**

  Processed foods typically lack dietary fiber, a critical nutrient for gut bacteria. Without sufficient fiber, beneficial bacteria populations decline, reducing SCFA production and impairing gut health. At the same time, harmful bacteria may thrive, producing endotoxins such as lipopolysaccharides (LPS) that further compromise the intestinal barrier and activate immune responses (Cani et al., 2009).

- **Inflammation and Autoimmune Triggers**

  Dysbiosis can lead to chronic low-grade inflammation, a hallmark of autoimmune disorders. Overactivation of immune cells in the gut can spill over into systemic circulation, affecting other organs and tissues. For instance:

  - In **rheumatoid arthritis**, gut inflammation is linked to joint inflammation, suggesting that microbiota imbalances may contribute to disease progression.

  - In **inflammatory bowel disease**, disruptions in the gut microbiome exacerbate intestinal inflammation, leading to severe symptoms and complications.

Additives and Immune Response

Processed foods are laden with additives designed to improve shelf life, flavor, and texture. However, many of these additives have unintended consequences for immune health.

- **Preservatives and Artificial Colors**

  Some preservatives, such as **sodium benzoate** and artificial colorings, have been shown to induce oxidative stress and inflammatory responses. While these effects are more pronounced at high doses, long-term exposure through frequent consumption of processed foods may contribute to immune dysregulation.

- **Advanced Glycation End Products (AGEs)**

  Processed foods often contain high levels of AGEs, compounds formed when sugars react with proteins or fats during high-temperature cooking. AGEs promote oxidative stress and inflammation, both of which are implicated in the development of autoimmune conditions.

Dietary Implications and Management

Addressing the role of processed foods in autoimmune disorders requires a focus on dietary patterns that support gut health and reduce inflammation.

- **Anti-Inflammatory Diets**

  Diets rich in anti-inflammatory foods, such as fruits, vegetables, whole grains, nuts, and fatty fish, can help mitigate the effects of gut dysbiosis and chronic inflammation. Incorporating **probiotic** and **prebiotic** foods, such as yogurt, kefir, and high-fiber vegetables, can restore microbial balance and strengthen the gut barrier.

- **Avoidance of Processed Foods**

  Reducing or eliminating processed foods from the diet can help limit exposure to harmful additives and promote overall gut health. For individuals with autoimmune conditions, this dietary change can reduce symptom severity and improve quality of life.

The link between processed foods and autoimmune disorders highlights the far-reaching impact of modern dietary habits on immune health. By compromising gut permeability, disrupting microbiota balance, and promoting chronic inflammation, processed foods contribute to the development and exacerbation of conditions like rheumatoid arthritis, lupus, and inflammatory bowel disease. A focus on whole, nutrient-dense foods and strategies to restore gut health offers a promising approach to managing autoimmune conditions and reducing their prevalence in an increasingly processed-food-dependent world.

How Inflammation is Triggered by Chemical Additives and Refined Ingredients

Chronic inflammation is a key mechanism by which processed foods contribute to the development of chronic diseases. This persistent, low-grade inflammatory state is driven by a combination of chemical additives, refined ingredients, and the oxidative stress induced by certain compounds in processed foods. These factors disrupt metabolic processes, immune functions, and cellular health, creating a foundation for conditions such as diabetes, cardiovascular diseases, and autoimmune disorders.

Chemical Additives and Inflammation

Chemical additives are ubiquitous in processed foods, serving roles such as improving texture, extending shelf life, and enhancing flavor. However, many of these substances have been shown to induce inflammatory responses, particularly in the gut.

- **Emulsifiers**

  Emulsifiers, such as carboxymethylcellulose and polysorbate-80, are commonly added to products like ice cream, dressings, and baked goods to stabilize mixtures and improve mouthfeel. Research has demonstrated that these additives can disrupt the gut microbiota, the community of microorganisms that plays a critical role in regulating immune and metabolic functions. By altering microbial composition, emulsifiers promote the proliferation of harmful bacteria that compromise the intestinal lining, leading to increased gut permeability (Chassaing et al., 2015). This condition, often referred to as "leaky gut," allows toxins and bacterial byproducts like lipopolysaccharides (LPS) to enter the bloodstream, triggering systemic inflammation and contributing to metabolic syndrome.

- **Preservatives and Artificial Colorings**

  Preservatives such as sodium benzoate and potassium sorbate are designed to inhibit microbial growth and prolong shelf life, but they can also generate reactive oxygen species (ROS) that exacerbate oxidative stress and inflammation. Artificial colorings, such as tartrazine (Yellow 5) and Allura Red (Red 40), have been implicated in hyperactivity and immune system dysregulation in some individuals. While regulatory agencies consider these additives safe at approved levels, their cumulative effects, particularly in diets dominated by processed foods, remain a concern.

Refined Ingredients

Refined ingredients, particularly sugars and oils, are central to the inflammatory profile of processed foods. These substances are often stripped of nutrients and fiber, creating metabolic imbalances that promote inflammation.

- **Refined Sugars and Carbohydrates**

  Refined sugars, such as high-fructose corn syrup (HFCS) and sucrose, are rapidly absorbed into the bloodstream, causing sharp increases in blood glucose and insulin levels. This glycemic spike leads to the production of pro-inflammatory cytokines such as interleukin-6 (IL-6) and tumor necrosis factor-alpha (TNF-α). Chronic exposure to these inflammatory mediators damages tissues, impairs insulin sensitivity, and increases the risk of type 2 diabetes and cardiovascular diseases.

In addition to their direct inflammatory effects, refined carbohydrates contribute to obesity, a condition characterized by the accumulation of visceral fat. Visceral fat is metabolically active and releases inflammatory molecules that exacerbate systemic inflammation.

- **Refined Oils and Omega-6 Fatty Acids**

  Many processed foods contain oils rich in omega-6 fatty acids, such as soybean, corn, and sunflower oils. While omega-6 fatty acids are essential in small amounts, their overconsumption relative to omega-3 fatty acids creates a pro-inflammatory environment. This imbalance promotes the production of eicosanoids, signaling molecules that drive inflammation and increase the risk of chronic diseases like arthritis, cardiovascular disease, and inflammatory bowel disease (Simopoulos, 2002).

Advanced Glycation End Products (AGEs)

Advanced glycation end products (AGEs) are harmful compounds formed during high-temperature cooking methods like grilling, frying, and roasting. These compounds result from a chemical reaction between sugars and proteins or fats, known as the Maillard reaction, which gives foods a desirable flavor and color.

- **Impact on Inflammation**

  AGEs bind to specific receptors on cell surfaces, known as RAGE (receptor for advanced glycation end products), activating inflammatory pathways. This activation increases oxidative stress, damages tissues, and contributes to the progression of diseases such as diabetes, cardiovascular disease, and kidney dysfunction.

- **Dietary Sources of AGEs**

  Processed foods such as fried snacks, baked goods, and processed meats are particularly high in AGEs. Regular consumption of these foods has been linked to elevated circulating AGE levels, which correlate with markers of systemic inflammation.

The Role of Processed Foods in the Global Rise of Non-Communicable Diseases

The increasing consumption of processed foods parallels the global epidemic of non-communicable diseases (NCDs), including diabetes, obesity, cardiovascular diseases, and cancers. These diseases are now the leading cause of death worldwide, accounting for more than 70% of global mortality. The dietary transition toward ultra-processed foods has fundamentally altered the nutritional landscape, exacerbating public health challenges.

Global Dietary Shifts

Economic globalization and urbanization have driven significant changes in dietary patterns, particularly in low- and middle-income countries. The widespread adoption of Western-style diets, characterized by high consumption of processed foods, sugary beverages, and refined ingredients, has replaced traditional diets rich in whole grains, legumes, and fresh produce.

- **Urbanization and Accessibility**

  Urban environments are increasingly dominated by fast food outlets, convenience stores, and supermarkets that prioritize ultra-processed products. These foods are often cheaper and more accessible than fresh alternatives, particularly in areas with limited agricultural infrastructure or food deserts.

- **Aggressive Marketing**

  Processed food companies invest heavily in marketing, often targeting vulnerable populations such as children and low-income families. Advertising campaigns promote the convenience, affordability, and taste of processed products while downplaying their negative health impacts.

- **Economic Drivers**

  Government subsidies for crops like corn, soy, and wheat lower the cost of ingredients used in processed foods, making them more affordable than nutrient-dense options. This economic incentive perpetuates dietary patterns that favor processed food consumption.

Health Implications

The global shift toward processed foods has profound implications for public health:

- **Diabetes and Obesity Epidemics**

  Diets dominated by processed foods contribute to excessive calorie intake, insulin resistance, and weight gain, fueling the rise of type 2 diabetes and obesity.

- **Cardiovascular Diseases**

  High levels of trans fats, refined sugars, and sodium in processed foods are directly linked to hypertension, atherosclerosis, and heart failure.

- **Cancer and Inflammatory Diseases**

  The carcinogenic compounds and pro-inflammatory nature of processed foods exacerbate the risk of cancers and other inflammatory conditions.

The inflammatory responses triggered by chemical additives, refined ingredients, and AGEs in processed foods illustrate their central role in the development of chronic diseases. Coupled with the global dietary shift toward ultra-processed products, these factors contribute to the escalating prevalence of NCDs worldwide. Addressing this crisis requires coordinated efforts to promote healthier dietary habits, regulate harmful food components, and improve access to nutrient-dense, whole foods.

Economic and Social Impacts

The consumption of processed foods is not only a health concern but also a driver of economic burdens. NCDs related to poor diet cost healthcare systems billions annually and reduce workforce productivity. Moreover, the aggressive marketing of processed foods, particularly to children and vulnerable populations, perpetuates unhealthy eating habits and exacerbates health disparities.

Policy and Public Health Interventions

Addressing the role of processed foods in NCDs requires comprehensive strategies, including:

- **Nutritional Labeling:** Front-of-package labels that highlight high levels of sugars, fats, and sodium can help consumers make healthier choices.

- **Taxation and Regulation:** Policies like sugar taxes and restrictions on trans fats have been effective in reducing consumption of unhealthy foods.

- **Education and Awareness Campaigns:** Public health initiatives aimed at promoting whole, minimally processed foods can empower individuals to adopt healthier diets.

Processed foods have become a staple of modern diets, but their convenience and affordability come at a significant cost to public health. From cancer and heart disease to autoimmune disorders, the links between processed foods and chronic diseases are undeniable. The mechanisms driving these connections, particularly through inflammation and metabolic disruption, highlight the urgent need for dietary shifts toward whole, nutrient-dense foods. By addressing the pervasive influence of processed foods, society can take critical steps toward reducing the burden of non-communicable diseases and fostering a healthier future.

Chapter 9

The Global Processed Food Economy

Processed foods are not merely a dietary choice but a global economic force that shapes supply chains, trade policies, and consumer behavior. The industrialization of food production has transformed how food is grown, processed, and distributed, creating a system that prioritizes efficiency and profitability. This chapter examines the intricate dynamics of the global processed food economy, focusing on the role of industrial food production, the influence of multinational food corporations, and the economic disparity between processed foods and fresh alternatives.

The industrialization of food production has revolutionized the way food is grown, processed, and delivered to consumers, creating a globalized system of supply chains that is both highly efficient and deeply interconnected. Enabled by technological advancements and transportation innovations, this system facilitates the mass production and distribution of processed foods on a scale that transcends geographic and economic boundaries. However, it also comes with significant implications for economic development, labor, and environmental sustainability.

Global Supply Chains and Processed Foods

Global supply chains are the backbone of industrial food production, enabling the sourcing of raw materials, incorporation of specialized additives, and widespread distribution of processed foods. These supply chains are highly complex and often span multiple continents, reflecting the integration of agriculture, manufacturing, and logistics on a global scale.

Staple Crops

Crops like **corn, soy,** and **wheat** are the foundation of processed food production. Heavily subsidized in many countries, these staples serve as the primary ingredients for a vast array of processed products, including snacks, beverages, and ready-to-eat meals.

- **Corn** is used to produce high-fructose corn syrup, a ubiquitous sweetener, as well as starches and feed for livestock.

- **Soy** provides oils, protein isolates, and emulsifiers, which are critical for textural consistency in processed foods.

- **Wheat** is the base for flours and gluten, essential for bread, pastries, and pasta.

The global trade of these crops creates interdependence between producing and consuming countries. For instance, the United States and Brazil dominate corn and soy production, while countries in Asia and Africa rely heavily on imports to meet the demands of their food industries.

Additives and Ingredients

Processed foods often include a variety of additives sourced from specialized suppliers worldwide. These include:

- **Preservatives** like sodium benzoate and potassium sorbate to extend shelf life.

- **Flavorings and Colorings** enhance sensory appeal.

- **Stabilizers and Emulsifiers** like lecithin and xanthan gum to improve texture and prevent separation in products like sauces and dairy alternatives.

The global nature of additive sourcing reflects the scale and complexity of modern food production, where components from different countries converge in manufacturing facilities to create a finished product.

Distribution Networks

Advances in transportation and logistics have enabled processed foods to reach consumers in both urban centers and remote regions. Key innovations include:

- **Cold Chain Technologies:** These systems maintain consistent temperature control throughout the supply chain, ensuring the quality and safety of perishable processed products like frozen meals and dairy items.

- **Modern Logistics:** Efficient shipping routes, advanced tracking systems, and economies of scale enable large volumes of processed foods to be transported globally at low cost. This extensive distribution infrastructure not only supports consumer access to processed foods but also allows manufacturers to expand their markets across borders, driving global food trade.

Economic Contributions

The processed food industry is a vital component of national and global economies, providing jobs, generating revenue, and spurring growth in related sectors. Industrial food production supports millions of jobs across the food supply chain, including:

- **Agriculture:** Farmers grow staple crops and supply raw materials essential for processing.

- **Manufacturing:** Factory workers oversee the transformation of raw ingredients into finished products.

- **Retail:** Supermarkets, convenience stores, and online platforms employ workers to distribute and sell processed foods. These employment opportunities are particularly significant in developing countries, where agricultural and manufacturing sectors form the backbone of the economy.

Revenue Generation

Processed food production and exports contribute significantly to national GDPs in major producing countries:

- The United States is a global leader in processed food exports, including snacks, cereals, and beverages.

- **Brazil** dominates in soy-based products and meat processing, driven by its vast agricultural resources.

- **China** has become a hub for producing and exporting processed foods, leveraging its advanced manufacturing capabilities and expansive supply chains. These revenues fuel economic growth and provide governments with the financial resources to invest in infrastructure, education, and healthcare.

Related Industries

The growth of industrial food production has spurred development in associated sectors, including:

- **Packaging:** Innovations in food-safe, eco-friendly materials have emerged to meet the demand for durable and attractive packaging.

- **Transportation:** Shipping companies benefit from the high volumes of goods moved across domestic and international routes.

- **Marketing:** Advertising and branding play a critical role in promoting processed foods, employing creative professionals and media outlets. These industries collectively contribute to a robust economic ecosystem, highlighting the interconnectivity of the processed food economy.

Challenges and Trade-Offs

While industrial food production has delivered economic benefits, it also poses challenges that require careful management:

- **Environmental Impact:** Monoculture farming practices associated with staple crops deplete soil nutrients, contribute to deforestation, and increase greenhouse gas emissions.

- **Economic Inequality:** Small-scale farmers and local food producers often struggle to compete with large industrial operations, leading to disparities in economic opportunity.

- **Supply Chain Vulnerabilities:** Global supply chains are susceptible to disruptions from geopolitical tensions, pandemics, and climate change, which can impact food availability and prices.

Industrial food production has reshaped the global food landscape, creating intricate supply chains that connect farmers, manufacturers, and consumers across continents. While this system has fueled economic growth and expanded access to food, it also presents challenges that demand sustainable solutions. Understanding the dynamics of global supply chains and their economic contributions is essential for fostering a food system that balances efficiency, equity, and environmental responsibility.

Environmental and Social Costs

The industrial food production system, while economically beneficial, comes with significant environmental and social costs. These costs undermine long-term sustainability and equity, raising questions about the true price of processed foods on the planet and society.

Environmental Costs

Monoculture farming, though efficient for large-scale food production, comes at a steep environmental cost, depleting soil, reducing biodiversity, and accelerating climate change.

Monoculture Farming Practices

Monoculture farming, the cultivation of a single crop over vast areas, is a cornerstone of industrial food production. While it enhances efficiency and simplifies harvesting, it poses severe environmental risks:

- **Soil Fertility Depletion:** Growing the same crop repeatedly depletes specific nutrients from the soil, reducing its fertility over time. This leads to increased reliance on synthetic fertilizers, which further degrade soil health.

- **Biodiversity Loss:** Monocultures eliminate diverse plant and animal species, disrupting ecosystems. The absence of natural predators also increases vulnerability to pests, necessitating the heavy use of pesticides.

- **Climate Impact:** Clearing land for large-scale monoculture contributes to deforestation and the release of carbon dioxide, exacerbating climate change.

High-Energy Processing Facilities

The facilities that process raw materials into finished products are energy-intensive, relying on fossil fuels to power machinery, refrigeration, and transportation:

- **Greenhouse Gas Emissions:** Processing facilities emit significant amounts of carbon dioxide (CO_2) and methane, contributing to global warming.

- **Water Usage:** Many facilities require vast amounts of water for cleaning, cooling, and processing, straining local water resources.

- **Packaging Waste:** The processed food industry generates immense quantities of single-use plastic and other non-biodegradable materials, much of which ends up in landfills or oceans, contributing to environmental degradation.

Social Costs

The social costs of processed foods extend far beyond the supermarket aisle—fueling public health crises, straining healthcare systems, and deepening inequalities, all while distancing communities from the true value of sustainable and nutritious eating.

Labor Conditions in Agricultural and Manufacturing Sectors

The industrial food system depends heavily on labor in both agriculture and manufacturing. However, these sectors are often criticized for exploitative practices:

- **Low Wages:** Workers in the agricultural supply chain, particularly in developing countries, often earn below living wages, perpetuating cycles of poverty.

- **Poor Working Conditions:** Harsh working conditions, lack of health benefits, and exposure to harmful chemicals are common in both farms and processing facilities. Seasonal workers and undocumented laborers are especially vulnerable to exploitation.

- **Child Labor and Exploitation:** In some regions, children are employed in agricultural supply chains, working long hours in hazardous conditions.

Community Displacement

The expansion of industrial agriculture often displaces small-scale farmers and indigenous communities, disrupting local food systems and eroding traditional practices. This displacement consolidates land ownership in the hands of large corporations, further marginalizing vulnerable populations.

The Rise of Multinational Food Corporations and Their Role in Dietary Changes Worldwide

Multinational corporations (MNCs) dominate the global processed food market, shaping dietary habits, trade policies, and consumer behavior. Companies like Nestlé, PepsiCo, Unilever, and Mondelez wield immense power, driving the globalization of food systems and influencing what billions of people eat.

Market Consolidation

The processed food industry is characterized by a high degree of market consolidation, where a few dominant players control vast portions of the market. This concentration of power has far-reaching implications for farmers, consumers, and policymakers.

Dictating Pricing and Sourcing Terms

MNCs leverage their scale to dictate prices and sourcing terms to farmers and suppliers:

- **Farmer Dependence:** Small-scale farmers often become dependent on selling their crops to these corporations, leaving them with little bargaining power. This dependency can drive down prices, making it difficult for farmers to earn sustainable incomes.

- **Global Supply Chains:** By controlling sourcing practices, MNCs can move production to regions with the lowest labor and material costs, maximizing profitability but often at the expense of local economies and workers' rights.

Shaping Consumer Preferences Through Marketing

MNCs invest billions in advertising and branding to influence consumer choices:

- **Targeted Advertising:** Marketing campaigns target specific demographics, including children, by promoting products as convenient, affordable, and aspirational. These efforts often overshadow the health risks associated with processed foods.

- **Cultural Influence:** Branded products are marketed as symbols of modernity and success, leading to shifts away from traditional diets, particularly in developing countries.

- **Product Placement:** MNCs secure prominent shelf space in supermarkets and convenience stores, ensuring that their products dominate consumer options.

Influencing Regulatory Frameworks and Trade Policies

The economic clout of multinational food corporations allows them to shape policies and regulations in their favor:

- **Lobbying Efforts:** MNCs exert influence on governments to delay or block regulations that would limit the use of harmful ingredients or mandate clear labeling of processed foods.

- **Trade Agreements:** These corporations play a significant role in shaping international trade agreements, prioritizing market access and tariff reductions for processed food exports over the needs of small-scale farmers and local food systems.

- **Undermining Public Health Policies:** In some cases, MNCs have been accused of undermining efforts to curb the consumption of unhealthy foods, such as opposing soda taxes or front-of-package warning labels.

The environmental and social costs of industrial food production, combined with the growing influence of multinational food corporations, underscore the complex dynamics of the global processed food economy. While these systems deliver efficiency and profitability, they also exacerbate environmental degradation, labor exploitation, and public health challenges. Addressing these issues requires coordinated efforts across governments, industries, and consumers to create a more equitable and sustainable food system.

Dietary Globalization

Multinational corporations play a pivotal role in spreading Western-style diets characterized by high consumption of ultra-processed foods:

- **Advertising:** Aggressive marketing campaigns target diverse demographics, including children and low-income populations, promoting products that are calorie-dense and nutrient-poor.

- **Product Availability:** By leveraging global distribution networks, MNCs ensure that processed foods are accessible in urban centers and rural communities worldwide.

- **Cultural Influence:** Branded foods often symbolize modernity and convenience, driving shifts away from traditional diets rich in whole grains, vegetables, and legumes.

Impacts on Public Health

The dominance of multinational food corporations has contributed to a global dietary transition associated with rising rates of obesity, diabetes, and cardiovascular diseases. The widespread availability of sugary beverages, salty snacks, and processed meals has exacerbated these public health challenges, particularly in low- and middle-income countries.

The Economics of Cheap, Processed Food vs. Fresh Alternatives

Processed foods are often significantly cheaper than fresh, whole foods, making them a staple for many households. This economic disparity arises from a combination of factors related to production, subsidies, and supply chain efficiencies.

Why Processed Foods Are Cheaper

- **Subsidized Ingredients:** Governments subsidize staple crops like corn, soy, and wheat, which are primary ingredients in processed foods. These subsidies lower production costs and allow manufacturers to price their products competitively.

- **Economies of Scale:** Large-scale production facilities and automated processes reduce the cost per unit, enabling manufacturers to offer low prices.

- **Long Shelf Life:** Processed foods are designed to have extended shelf lives, reducing waste and inventory costs for retailers.

Why Fresh Foods Are Expensive

- **Perishability:** Fresh produce and meat have short shelf lives, requiring specialized storage and transportation systems that increase costs.

- **Labor-Intensive Production:** Growing, harvesting, and distributing fresh foods involve higher labor inputs compared to mechanized processing facilities.

- **Seasonal Variability:** Fresh foods are subject to price fluctuations due to seasonal availability and climate conditions.

Consumer Implications

The affordability of processed foods disproportionately affects low-income households, where budget constraints drive purchasing decisions. For these consumers, processed foods offer a cost-effective way to meet caloric needs, albeit at the expense of nutritional quality. This economic reality exacerbates health disparities, as wealthier individuals are better positioned to access fresh, nutrient-dense foods.

The global processed food economy is a complex system that shapes supply chains, consumer diets, and economic structures. While it has delivered significant benefits in terms of efficiency and affordability, it has also contributed to public health challenges and environmental degradation. Understanding the dynamics of industrial food production, the influence of multinational corporations, and the economic disparities between processed and fresh foods is essential for creating a more equitable and sustainable global food system.

Chapter 10

Environmental Cost of Processed Foods

The environmental cost of processed foods extends far beyond the factory walls, infiltrating every stage of production, distribution, and disposal. From the significant carbon emissions generated by industrialized food systems to the destruction of ecosystems through deforestation and water pollution, the hidden price of convenience and shelf stability is borne by the planet. Compounding these impacts is the overwhelming volume of packaging waste that chokes landfills and marine environments, illustrating the profound and lasting imprint of processed foods on the Earth's ecosystems. Understanding these environmental consequences is critical to fostering more sustainable food production and consumption practices.

The Carbon Footprint of Industrialized Food Production

The carbon footprint of processed foods is a significant contributor to global greenhouse gas emissions, a key driver of climate change. This impact is rooted in the energy-intensive nature of the industrialized food production process, which encompasses several stages, including the extraction of raw materials, the manufacturing of food products, transportation to distribution centers and retailers, and ultimately, the energy needed for storage and preparation by consumers. Each of these stages relies heavily on fossil fuels, which release carbon dioxide (CO_2) and other greenhouse gases (GHGs) into the atmosphere.

For instance, the production of ready-to-eat meals involves multiple steps that are particularly energy-intensive. Ingredients are transported from farms to processing plants, where they undergo freezing, cooking, drying, or other forms of preparation, all of which require substantial energy inputs. The packaging process, often involving materials like plastics, aluminum, and cardboard, further compounds the energy demand. Additionally, the refrigerated transportation of frozen or chilled processed foods contributes significant emissions, as refrigeration systems often leak hydrofluorocarbons (HFCs), potent greenhouse gases with a global warming potential far exceeding that of CO_2 (Garnett, 2011).

Food processing plants themselves are often major sources of GHGs. In addition to CO_2 emissions from energy use, many facilities emit methane (CH_4) and nitrous oxide (N_2O) during production. Methane, released from food waste and wastewater treatment systems, has a global warming potential over 25 times greater than CO_2 over a 100-year period. Nitrous oxide, commonly associated with the use of fertilizers in agricultural inputs for processed foods, has an even higher warming potential, approximately 300 times that of CO_2 (IPCC, 2021).

The cumulative effect of these emissions underscores the substantial environmental toll of processed foods. Studies have demonstrated that diets emphasizing minimally processed or whole foods not only improve human health but also significantly reduce carbon emissions. Clark et al. (2019) found that shifting dietary patterns toward plant-based and less-processed options could decrease GHG emissions from the food system by as much as 49%. This highlights the potential for both individual and systemic changes to reduce the carbon footprint of food production.

Ultimately, addressing the carbon footprint of industrialized food production requires coordinated efforts at multiple levels. Governments can implement policies that incentivize sustainable practices, such as reducing energy use in processing plants or investing in renewable energy. Food companies can innovate in product design, emphasizing less energy-intensive processes and materials. Finally, consumers can make impactful choices by favoring fresh, local, and minimally processed foods. Together, these efforts can help mitigate the environmental impact of the global food system.

Food Processing Contributes to Water Pollution, Deforestation, and Biodiversity Loss

The environmental footprint of food processing extends far beyond carbon emissions; it also contributes significantly to water pollution, deforestation, and biodiversity loss. These impacts, though often overlooked, are interconnected and exacerbate the environmental crises facing the planet.

Water Pollution

Food processing industries are among the largest contributors to water pollution globally. Processing plants often discharge untreated or inadequately treated effluents containing harmful substances such as fats, oils, salts, and chemical additives into nearby water bodies. These pollutants degrade water quality, leading to increased biochemical oxygen demand (BOD) and decreased oxygen levels in aquatic ecosystems, which harms fish, plants, and microorganisms. Moreover, wastewater from food processing facilities frequently contains nitrogen and phosphorus compounds, which contribute to eutrophication—a process that causes algal blooms and "dead zones" where aquatic life cannot survive (Kumar & Sharma, 2020).

For example, meat processing plants are notorious for releasing effluents rich in organic matter and pathogens, which not only threaten aquatic ecosystems but also pose significant risks to human health. Similarly, dairy processing facilities often discharge high levels of fats and proteins, which can clog waterways and disrupt aquatic ecosystems. As global demand for processed foods continues to rise, so too does the burden on freshwater resources.

Deforestation

The demand for raw materials used in processed foods has driven widespread deforestation, particularly in tropical regions. Ingredients such as palm oil and soybeans, staples in processed food production, are often cultivated in plantations established through the clearing of forests. For instance, palm oil plantations are a leading cause of deforestation in Southeast Asia, with millions of hectares of forest converted to agricultural land (Wilcove & Koh, 2010). This deforestation not only releases vast amounts of carbon dioxide stored in trees but also destroys habitats critical to biodiversity.

In addition, the cultivation of soybeans—used in processed foods as an emulsifier, protein source, or animal feed—has led to deforestation in the Amazon rainforest. The loss of these forests accelerates global warming, disrupts regional water cycles, and drives countless species toward extinction.

Biodiversity Loss

The consequences of deforestation for biodiversity are severe and far-reaching. Forests are home to an estimated 80% of terrestrial species, many of which cannot survive outside their natural habitats. The destruction of these ecosystems fragments populations, reduces genetic diversity, and increases the vulnerability of species to extinction. Iconic species such as orangutans in Borneo and Sumatra have seen their populations plummet due to habitat loss driven by palm oil plantations (Meijaard et al., 2018).

Beyond deforestation, industrialized agriculture tied to processed foods exacerbates biodiversity loss through monoculture farming practices. These practices reduce soil fertility, increase susceptibility to pests and diseases, and threaten pollinators such as bees, which are essential for the reproduction of many crops. The widespread use of pesticides and herbicides further harms pollinators and other non-target species, compounding the biodiversity crisis.

Interconnected Impacts

The relationship between water pollution, deforestation, and biodiversity loss is deeply interconnected. Pollutants from food processing often enter ecosystems already weakened by deforestation, compounding the stress on biodiversity. Additionally, the loss of forests disrupts water cycles, reducing the availability of clean water for ecosystems and human communities alike. These cascading impacts illustrate the complex and pervasive environmental costs of industrialized food systems.

Addressing these challenges requires coordinated action across the supply chain. Governments must enforce stricter regulations on effluent treatment and incentivize sustainable agricultural practices. Companies must adopt transparency and accountability in sourcing raw materials, avoiding ingredients linked to deforestation or habitat destruction. Finally, consumers can play a role by choosing products that prioritize sustainability, such as those certified by organizations like the Rainforest Alliance or Fair Trade.

Packaging Waste and Its Impact on Ecosystems

The packaging of processed foods is a major contributor to global environmental pollution, with significant and long-lasting impacts on ecosystems. Processed food packaging, often made from materials such as plastic, aluminum, and multilayer composites, is designed for durability and convenience. However, these same properties make it a significant environmental hazard, as most packaging is non-biodegradable and persists in the environment for decades or even centuries.

The Proliferation of Single-Use Plastics

Single-use plastics are a dominant material in processed food packaging due to their low cost, lightweight nature, and ability to preserve food quality. However, their environmental footprint is staggering. Plastics take hundreds of years to decompose, during which they break down into microplastics—tiny fragments that contaminate soil, waterways, and marine environments. Each year, an estimated 8 million metric tons of plastic waste enter the oceans, much of it originating from food packaging (Jambeck et al., 2015). This waste forms massive garbage patches, such as the Great Pacific Garbage Patch, which spans thousands of square miles and poses a direct threat to marine life.

Impact on Wildlife and Ecosystems

Marine and terrestrial ecosystems bear the brunt of the environmental toll from packaging waste. Animals often mistake plastic waste for food, leading to ingestion that causes choking, internal injuries, or starvation due to blocked digestive tracts. For example, sea turtles commonly consume plastic bags, mistaking them for jellyfish, while seabirds ingest bottle caps and other fragments (Gall & Thompson, 2015). Additionally, microplastics have been found in the tissues of marine organisms, entering the food chain and potentially impacting human health.

Packaging waste not only harms individual species but also disrupts entire ecosystems. In marine environments, floating plastics provide surfaces for invasive species to travel and colonize new areas, altering ecosystem dynamics. On land, improperly managed landfills leak packaging materials into nearby ecosystems, contaminating soils and groundwater with chemicals leached from plastics and metals.

Challenges of Recycling and Energy Use

Despite global awareness of the problem, recycling rates for processed food packaging remain dismally low. Complex packaging designs, such as multi-layer materials that combine plastic and aluminum, are often non-recyclable due to the difficulty of separating the layers. Even recyclable materials like aluminum and PET plastics require significant energy inputs for processing, compounding their environmental footprint. Moreover, the production of packaging materials is itself energy-intensive, relying on fossil fuels and contributing to greenhouse gas emissions.

Addressing the Problem

Addressing the environmental impact of processed food packaging requires systemic changes in production, consumption, and waste management. Governments can implement policies to phase out single-use plastics, encourage the development of biodegradable alternatives, and improve recycling infrastructure. The food industry must innovate in packaging design, prioritizing materials that are both sustainable and functional. For example, plant-based biodegradable materials or reusable packaging systems can reduce environmental impacts. Consumers also play a crucial role by choosing products with minimal or recyclable packaging and supporting brands that prioritize sustainability.

The environmental challenges posed by processed food packaging highlight the urgent need for a shift toward a circular economy, where materials are reused and recycled efficiently. By addressing packaging waste, society can mitigate one of the most visible and pervasive forms of environmental degradation.

The environmental cost of processed foods is a stark reminder that the convenience of modern diets comes at a steep price to our planet. The carbon emissions, deforestation, water pollution, and packaging waste associated with these foods are not just abstract issues—they are urgent challenges that threaten ecosystems, biodiversity, and the future of life on Earth. As consumers, policymakers, and industries, we have the power to demand and implement changes that prioritize sustainability over profit and convenience. The choices we make today will determine whether we leave behind a legacy of degradation or one of stewardship and resilience for future generations.

Chapter 11

Processed Foods and Food Inequality

Food inequality is a pressing issue that highlights the stark disparities in access to nutritious food, particularly between affluent and low-income communities. Processed foods, often more affordable and accessible than fresh, wholesome options, dominate the diets of those in underserved areas, perpetuating cycles of poor health and economic disadvantage. The prevalence of food deserts, coupled with targeted marketing that exploits vulnerable populations, underscores the systemic nature of this problem. Understanding how processed foods contribute to food inequality is essential to addressing the broader challenges of health disparities and social justice in the modern food system.

Why Processed Foods Are More Accessible Than Fresh Foods in Many Areas

The widespread accessibility of processed foods over fresh alternatives stems from the design and priorities of industrialized food systems, shaped by economic incentives and disparities in infrastructure. These systems are structured to prioritize long shelf life, cost efficiency, and mass distribution, characteristics that favor processed foods. In contrast, fresh foods, which require refrigeration, frequent replenishment, and shorter supply chains, are inherently more challenging to make widely available, particularly in underserved areas.

The Role of Industrialized Food Systems

Processed foods dominate the global market because they are engineered for stability and convenience. Products like canned soups, frozen meals, and snack foods can remain on store shelves for months or even years without spoiling. This durability reduces losses for retailers and ensures steady availability for consumers, even in areas with limited infrastructure. Additionally, processed foods are compact and lightweight, making them easier and cheaper to transport over long distances compared to fresh produce, which is bulkier, heavier, and more perishable.

The convenience of processed foods also extends to their preparation. Many processed products are ready-to-eat or require minimal effort, catering to the fast-paced lifestyles of modern consumers. In contrast, fresh foods often require time, effort, and knowledge to prepare, factors that can deter consumers, especially those with demanding schedules or limited cooking skills.

Economic Incentives and Government Subsidies

Economic policies further widen the gap between processed and fresh foods. In the United States, government subsidies disproportionately benefit commodity crops such as corn, wheat, and soybeans. These crops are the building blocks of processed foods, providing inexpensive ingredients like high-fructose corn syrup, vegetable oils, and fillers (Okrent & Alston, 2011). The subsidies lower production costs for food manufacturers, enabling them to offer processed foods at prices far below those of fresh fruits and vegetables, which receive minimal government support.

For example, a bag of chips made from subsidized corn may cost less than a fresh apple, despite the former undergoing multiple stages of processing. This price disparity is particularly impactful in low-income communities, where families are often forced to prioritize calories per dollar over nutritional value.

Infrastructure and Distribution Challenges

The infrastructure supporting food distribution is another critical factor contributing to the accessibility of processed foods. Processed products require minimal storage demands, as they do not need refrigeration and are resistant to spoilage. This makes them ideal for distribution to remote or underserved areas where cold storage facilities are scarce or non-existent.

In contrast, fresh foods have far more stringent storage and transportation requirements. They must be kept at specific temperatures to maintain freshness, necessitating refrigerated trucks and cold storage facilities along the supply chain. Such infrastructure is expensive to build and maintain, and its scarcity in low-income and rural areas limits the availability of fresh produce. Additionally, the shorter shelf life of fresh foods increases the risk of spoilage, leading to higher costs for retailers and, ultimately, consumers.

Impact on Underserved Areas

In low-income and rural areas, these disparities are particularly pronounced. Processed foods, which are cheaper to produce and easier to store, dominate the shelves of convenience stores and small markets that lack the resources to stock fresh produce. Supermarkets with extensive fresh food offerings are often located in wealthier neighborhoods, leaving residents of underserved areas reliant on less nutritious options.

This accessibility gap not only perpetuates poor dietary habits but also entrenches health disparities. Diets high in processed foods are linked to chronic diseases such as obesity, diabetes, and cardiovascular conditions, which disproportionately affect low-income populations. The convenience and affordability of processed foods thus create a paradox: they are readily accessible, yet their long-term impacts contribute to the very inequalities they seem to alleviate.

Potential Solutions

Addressing the accessibility of fresh foods requires systemic changes. Policies that redirect subsidies toward fruits and vegetables could help lower their cost and increase their availability. Investments in cold storage infrastructure, particularly in rural and low-income areas, would enable better distribution of fresh produce. Community-based solutions, such as farmers' markets, urban agriculture, and food co-ops, can also bridge the gap by bringing fresh foods closer to underserved populations.

Ultimately, addressing this imbalance is not just about making fresh foods more accessible but about rethinking the systems and policies that prioritize processed foods over nutrition and equity. By realigning economic and infrastructural priorities, it is possible to create a food system that better serves the health and well-being of all communities.

The Impact on Low-Income Communities and Food Deserts

The disparity in access to fresh and processed foods has profound consequences for low-income communities, particularly those situated in food deserts. These are areas where affordable and nutritious food is scarce, often leaving residents reliant on convenience stores and fast-food outlets stocked predominantly with processed and ultra-processed foods. This reliance has dire health implications, as diets rich in processed foods are strongly linked to increased risks of obesity, type 2 diabetes, and cardiovascular disease (Micha et al., 2017).

Health Implications of Processed Foods

In food deserts, the lack of access to fresh produce and whole foods forces residents to prioritize affordability and caloric density over nutritional value. Processed foods, which are high in added sugars, unhealthy fats, and sodium, are often the most cost-effective option. For low-income families, these calorie-dense foods may appear to offer the best return on investment, but the long-term consequences are severe. High consumption of processed foods has been linked to rising rates of diet-related diseases, disproportionately affecting marginalized communities that already face barriers to healthcare access.

The burden of poor nutrition extends beyond individual health, impacting families and communities as a whole. Increased medical expenses, reduced productivity, and shortened lifesp

ans contribute to a cycle of poverty and poor health, further entrenching the challenges faced by low-income populations. This dynamic illustrates how food inequality exacerbates broader social and economic disparities.

Food Deserts as a Symptom of Systemic Inequality

Food deserts are not merely the result of chance or market forces; they are deeply rooted in systemic inequality. Urban planning decisions often prioritize affluent neighborhoods, resulting in an uneven distribution of grocery stores and food outlets. Full-service grocery stores, which offer fresh produce and healthier options, are more likely to be located in wealthier areas, while low-income neighborhoods are left with convenience stores or small markets that predominantly stock processed foods.

Economic policies and zoning laws also play a role. Many low-income neighborhoods lack incentives for grocery chains to establish stores, due to perceived low profit margins or concerns about safety and infrastructure. This creates a vicious cycle: without access to nutritious food options, residents are forced to rely on what is available, perpetuating poor health outcomes and reduced demand for healthier options.

Transportation Barriers and Accessibility Challenges

Transportation barriers compound the issue, particularly for residents of rural areas or urban neighborhoods poorly served by public transit.

Without reliable access to transportation, residents may struggle to reach stores that offer fresh and nutritious food. For families without personal vehicles, the effort and expense of traveling long distances for groceries can be prohibitive. This logistical challenge effectively traps individuals in food deserts, reinforcing the dominance of processed foods in their diets (Walker et al., 2010).

Breaking the Cycle

Addressing the impact of food deserts on low-income communities requires systemic interventions. Governments and community organizations can play a pivotal role in mitigating food inequality through policies and programs designed to increase access to nutritious foods. Potential strategies include:

- **Incentivizing Grocery Stores**: Offering tax breaks or subsidies to grocery chains and farmers' markets to establish operations in underserved areas.

- **Community-Based Solutions**: Encouraging urban agriculture, food co-ops, and mobile food markets to provide fresh produce directly to residents.

- **Improved Transportation Infrastructure**: Expanding public transit systems to connect food desert communities with grocery stores and farmers' markets.

Ultimately, breaking the cycle of food deserts and poor health requires a multifaceted approach that addresses both the systemic roots of inequality and the immediate needs of affected communities. By prioritizing equity in food access, we can create healthier and more resilient communities.

How Marketing and Advertising Target Vulnerable Populations

Processed food companies employ aggressive and strategic marketing tactics that disproportionately affect vulnerable populations, including low-income families, children, and minorities. These tactics exploit the socioeconomic vulnerabilities of these groups, using messaging that emphasizes affordability, convenience, and taste to promote processed foods. For many low-income families, these advertisements appeal directly to their lived realities, where time and financial constraints often make healthier options seem unattainable. Fast-food chains and snack companies, for example, frequently concentrate their marketing efforts in minority-dominated neighborhoods, saturating these areas with billboards, storefront signage, and targeted promotions. Similarly, processed food advertisements dominate television programming popular among children and teenagers, further shaping dietary preferences at an early age (Harris et al., 2019).

The Role of Digital Advertising

The rise of digital advertising has amplified the impact of these strategies. Social media platforms and search engines use sophisticated algorithms to target specific demographics based on income, geographic location, and online behavior. This allows processed food companies to deliver tailored advertisements directly to those most likely to buy their products. For instance, families with lower incomes may see more ads for fast food or prepackaged meals that emphasize low prices and convenience, while healthier food options remain underrepresented in these campaigns.

Children are particularly vulnerable to both traditional and digital marketing. In grocery stores, for example, sugary cereals are often placed at the lowest shelves, right at eye level for young children. This strategic placement is designed to catch their attention and encourage them to ask their parents to purchase these cereals, often over healthier options that are placed higher up and less accessible. This "pester power" tactic is highly effective, as children exert considerable influence over their parents' purchasing decisions. Research shows that food advertising directed at youth overwhelmingly promotes products high in sugar, salt, and unhealthy fats, contributing to poor dietary habits and long-term health risks (Powell et al., 2013).

Reinforcing Cultural Norms and Undermining Public Health

Marketing strategies for processed foods often go beyond simply selling a product; they shape cultural norms and perceptions of food. Advertisements frequently associate processed foods with convenience, modernity, and social status, creating a narrative that aligns these products with aspirational lifestyles. For instance, campaigns may portray fast food as a solution for busy families or highlight the indulgence of sugary snacks as a deserved reward. These messages reinforce the idea that processed foods are not only practical but desirable, overshadowing public health initiatives aimed at promoting healthier eating habits.

In many cases, these marketing efforts directly conflict with growing awareness of the negative health impacts of processed foods. Even as public health campaigns stress the importance of reducing sugar, sodium, and unhealthy fats, processed food advertisements continue to dominate the media landscape, undermining efforts to promote dietary change. The strategic placement of unhealthy food products, combined with pervasive advertising, creates an environment where healthier choices are harder to make, especially for families under economic or social pressure.

The pervasive presence of processed foods in low-income communities and food deserts is not just a symptom of inequality—it is a driver of it. The affordability and accessibility of these foods, coupled with aggressive marketing tactics, perpetuate cycles of poor health, economic hardship, and limited opportunities for change. Addressing food inequality requires systemic reform, from reshaping agricultural policies and improving infrastructure to enforcing ethical marketing practices and empowering communities with education and resources. By confronting these challenges, we can move toward a more equitable food system that prioritizes health, dignity, and sustainability for all.

Chapter 12

The Science of Whole Foods

Whole foods, celebrated for their simplicity and nutritional integrity, are the foundation of a healthy diet. Unlike processed foods, which often lose vital nutrients through industrial manipulation, whole foods retain their natural composition, delivering a wealth of vitamins, minerals, fiber, and antioxidants essential for optimal health. From vibrant fruits and vegetables to nutrient-dense whole grains, these foods offer unparalleled benefits, supporting disease prevention, gut health, and overall well-being. By understanding the science behind whole foods and their natural chemistry, we can make informed choices that nourish our bodies and contribute to long-term health. This chapter explores the unique advantages of minimally processed and whole foods, delving into their biochemical richness and the essential role of antioxidants, fiber, and micronutrients in maintaining a balanced and thriving body.

How Minimally Processed and Whole Foods Benefit Our Bodies

Whole foods and minimally processed foods are foundational to a healthy diet, offering unparalleled nutritional benefits due to their nutrient density, natural composition, and absence of harmful additives. These foods remain as close to their natural state as possible, retaining essential vitamins, minerals, fiber, and bioactive compounds that are often stripped away during processing. By prioritizing whole foods, individuals can nourish their bodies with the tools necessary for optimal function and disease prevention, making these foods critical to long-term health and well-being.

Nutritional Superiority and Bioavailability

Unlike processed foods, which are frequently altered to extend shelf life or enhance flavor, whole foods maintain their original nutrient composition. This preservation ensures that essential nutrients such as vitamins, minerals, and antioxidants are delivered in forms that are more bioavailable—easier for the body to absorb and utilize effectively. For example, the vitamin C in oranges or the potassium in bananas is accompanied by other natural compounds that enhance their absorption and effectiveness in the body (Slavin, 2013). In contrast, processed foods often include synthetic additives that may not provide the same level of bioavailability, potentially limiting their health benefits.

Whole foods also offer a rich supply of macronutrients—proteins, carbohydrates, and fats—in their most natural and beneficial forms. Proteins from legumes and whole grains, carbohydrates from fruits and vegetables, and healthy fats from nuts and seeds are all superior to the refined and unhealthy versions found in processed alternatives. This balance of nutrients provides sustained energy, supports cellular repair, and aids in the maintenance of vital body functions.

Prevention of Chronic Diseases

One of the most significant benefits of whole foods is their role in preventing chronic diseases, which are a leading cause of morbidity and mortality worldwide. Diets rich in whole foods have been shown to reduce the risk of conditions such as obesity, type 2 diabetes, cardiovascular disease, and certain cancers. These protective effects are largely attributed to the natural fiber, antioxidants, and phytochemicals found in fruits, vegetables, and whole grains.

For instance, a study by Hu et al. (2014) demonstrated that individuals who consumed diets high in whole foods, such as leafy greens, berries, and whole grains, experienced a significantly lower risk of cardiovascular disease. The fiber in these foods helps to regulate blood sugar levels, reduce cholesterol, and support healthy blood pressure, all of which are critical for heart health. Similarly, the antioxidants and phytochemicals in whole foods combat oxidative stress and inflammation, key contributors to the development of chronic diseases.

Unlike processed foods, which are often loaded with added sugars, unhealthy fats, and sodium, whole foods work to reduce inflammation and support metabolic health. Chronic inflammation, often exacerbated by a diet high in processed foods, is linked to conditions such as arthritis, Alzheimer's disease, and certain types of cancer. By incorporating whole foods into daily meals, individuals can mitigate these risks and promote a healthier internal environment.

Support for Gut Health and the Microbiome

The gut plays a central role in overall health, influencing not only digestion but also immunity, mental health, and metabolic regulation. Whole foods are essential for maintaining a healthy gut microbiome, the diverse community of bacteria and microorganisms that reside in the digestive tract. The fiber in fruits, vegetables, and whole grains acts as a prebiotic, feeding the beneficial bacteria in the gut and fostering a balanced microbial environment.

A balanced microbiome has far-reaching benefits, including enhanced digestion, better nutrient absorption, and protection against harmful pathogens. In addition, the gut microbiome is intricately connected to the gut-brain axis, a communication pathway that links the gut to the brain. Research shows that a healthy microbiome, supported by fiber-rich whole foods, can positively influence mood, reduce symptoms of depression and anxiety, and improve cognitive function (Deehan et al., 2017).

Furthermore, whole foods contribute to gut health by reducing the intake of harmful substances found in processed foods, such as artificial sweeteners and emulsifiers, which have been shown to disrupt the gut microbiome. By choosing whole foods, individuals can promote a thriving gut environment that supports overall health and resilience.

Understanding the Natural Chemistry of Fruits, Vegetables, and Whole Grains

The natural chemistry of whole foods is fundamental to their role in promoting health and preventing disease. Fruits, vegetables, and whole grains are composed of a rich array of bioactive compounds, each contributing to their unique nutritional value. These compounds— carbohydrates, proteins, lipids, vitamins, minerals, and phytochemicals—work together synergistically, amplifying their positive effects on the body.

The Biochemical Composition of Fruits and Vegetables

Fruits and vegetables are nutritional powerhouses due to their abundance of water-soluble and fat-soluble vitamins, minerals, and phytochemicals. Water-soluble vitamins, such as vitamin C, play a crucial role in protecting the body from oxidative stress by neutralizing free radicals, unstable molecules that can damage cells and accelerate aging. Fat-soluble vitamins like vitamin E, found in avocados and spinach, provide additional antioxidant protection, supporting skin health, cellular repair, and immune function.

Phytochemicals are another key component of fruits and vegetables, contributing to their vibrant colors and potent health benefits. Carotenoids, the pigments responsible for the orange and yellow hues of carrots and sweet potatoes, are well-known for supporting eye health and reducing the risk of macular degeneration. Flavonoids, found abundantly in berries, onions, and citrus fruits, have been shown to improve cardiovascular health by reducing inflammation, lowering blood pressure, and enhancing vascular function (Liu, 2013).

Polyphenols, present in foods like apples and green tea, exhibit anticancer properties by inhibiting tumor growth and protecting DNA from oxidative damage.

The synergy between these bioactive compounds is particularly noteworthy. For example, the presence of vitamin C enhances the absorption of non-heme iron from plant sources, improving its bioavailability. Similarly, the combination of fiber and antioxidants in fruits and vegetables contributes to both gut health and systemic inflammation reduction, highlighting the interconnected benefits of their natural chemistry.

The Unique Composition of Whole Grains

Whole grains offer a unique nutritional profile that distinguishes them from their refined counterparts. Unlike refined grains, which are stripped of their bran and germ during processing, whole grains retain these nutrient-dense components, providing a wealth of health benefits. The bran, rich in dietary fiber, aids in digestion and promotes a healthy gut microbiome, while the germ is a concentrated source of vitamins, minerals, and healthy fats.

The complex carbohydrate structure of whole grains is another critical factor in their nutritional value. Starches and fibers in whole grains are digested more slowly than the simple carbohydrates found in processed grains, resulting in a gradual release of glucose into the bloodstream. This slow digestion helps regulate blood sugar levels, reducing the risk of insulin resistance and type 2 diabetes. Additionally, the fiber in whole grains provides a feeling of fullness, aiding in weight management and appetite control.

Whole grains are also a rich source of essential micronutrients. Magnesium in grains like quinoa and brown rice supports bone health and enzymatic reactions, while zinc in oats and barley plays a vital role in immune function and wound healing. Selenium, a powerful antioxidant found in whole wheat, contributes to thyroid health and protects cells from oxidative damage. These nutrients are preserved in whole grains, making them far superior to refined grains, which lose much of their nutritional value during processing.

The Synergy of Whole Foods' Chemistry

The health benefits of whole foods extend beyond the sum of their individual components due to the synergistic interactions among their nutrients. For example, the combination of fiber, antioxidants, and polyphenols in whole grains not only promotes heart health but also reduces systemic inflammation, a common precursor to chronic diseases. Similarly, the natural pairing of water-soluble and fat-soluble vitamins in fruits and vegetables enhances their absorption and bioavailability, maximizing their impact on cellular repair and immune function.

This synergy underscores the importance of consuming whole foods in their natural state. While supplements and fortified foods can provide isolated nutrients, they often lack the complex matrix of compounds found in whole foods, which work together to optimize health. By understanding and appreciating the natural chemistry of fruits, vegetables, and whole grains, individuals can make informed dietary choices that support long-term health and well-being.

The Role of Antioxidants, Fiber, and Natural Micronutrients

Antioxidants, fiber, and natural micronutrients are the cornerstone of whole foods' health benefits. These components work together to combat oxidative stress, regulate metabolism, and maintain overall physiological balance.

Antioxidants are compounds that neutralize free radicals, unstable molecules that can damage cells and contribute to aging and chronic diseases. Whole foods are a rich source of antioxidants, including vitamins C and E, selenium, and polyphenols. Studies have shown that diets high in antioxidant-rich foods, such as fruits and vegetables, reduce the risk of chronic diseases like cancer and cardiovascular disorders. For example, the polyphenols in green tea and dark chocolate improve heart health by enhancing endothelial function and reducing LDL cholesterol oxidation (Kris-Etherton et al., 2013).

Fiber, a type of carbohydrate found in plant-based foods, is essential for digestive health and metabolic regulation. Soluble fiber, found in oats and beans, helps lower cholesterol levels by binding to bile acids and preventing their absorption in the intestine. Insoluble fiber, present in whole grains and vegetables, adds bulk to stool and prevents constipation. Both types of fiber contribute to satiety, making it easier to maintain a healthy weight.

Natural micronutrients, including vitamins and minerals, are critical for bodily functions. Potassium in bananas and leafy greens helps regulate blood pressure, while calcium in broccoli and almonds supports bone health. Iron in whole grains and legumes is vital for oxygen transport in the blood, and zinc in seeds and nuts plays a role in immune function and wound healing. These micronutrients are better absorbed and utilized when consumed in their natural forms within whole foods, as opposed to synthetic supplements (Heaney, 2001).

The science of whole foods underscores their unparalleled ability to nourish the body, prevent disease, and promote overall health. Their natural composition, rich in antioxidants, fiber, and essential micronutrients, offers synergistic benefits that processed foods cannot replicate. Understanding and prioritizing whole foods in our diets is not just a matter of personal health—it is a step toward a more sustainable and balanced approach to nutrition and well-being.

Whole foods represent the essence of what it means to eat for health, vitality, and sustainability. Their unaltered, nutrient-rich composition provides the building blocks our bodies need to thrive, while their natural synergy fosters optimal physiological function and disease prevention. By embracing minimally processed and whole foods, we not only enhance our personal well-being but also contribute to a food culture that prioritizes nutrition over convenience. Understanding the science behind these foods empowers us to make choices that support a healthier, more balanced life—one that is rooted in the timeless wisdom of nature's bounty.

Chapter 13

Navigating Labels and Marketing Myths

In a world flooded with food choices, deciphering the truth behind packaging and marketing claims has become an essential skill for health-conscious consumers. Food labels, ingredient lists, and marketing slogans often present a façade of healthfulness, masking the reality of processed products laden with added sugars, unhealthy fats, and artificial additives. Greenwashing tactics and the pervasive "health halo" effect further complicate decision-making, leading consumers to perceive certain products as healthier or more sustainable than they truly are. This chapter delves into the complexities of food labeling, exposing the myths perpetuated by misleading marketing and offering practical strategies for identifying genuinely healthier options in the ever-expanding sea of processed choices.

Decoding Ingredient Lists and Nutritional Labels

The first step in navigating the complex world of food packaging is understanding how to decode ingredient lists and nutritional labels. Food labels are designed to provide consumers with essential information about what they are eating, yet they are often deliberately complicated or misleading. The ingredient list, typically presented in descending order by weight, can reveal the true composition of a product. Ingredients like added sugars, sodium, and artificial additives, if listed among the first few items, signal a product that may not be as healthy as marketing claims suggest (Campos et al., 2011).

For example, added sugars may be disguised under a variety of names, such as high-fructose corn syrup, dextrose, or cane sugar, making it difficult for consumers to identify them at a glance. Similarly, terms like "natural flavors" or "spices" often lack specificity, providing little transparency about their origins or potential health implications. Understanding the difference between whole ingredients (e.g., whole-grain wheat) and their refined counterparts (e.g., enriched flour) is crucial in evaluating the nutritional value of a product.

The nutritional facts panel provides quantitative data about a product's caloric content, macronutrients, and micronutrients. Key elements to assess include:

Serving Size-Misleading Standards: Serving size, a seemingly straightforward metric on nutritional labels, is often manipulated by manufacturers to present their products in a more favorable light. By defining unrealistically small serving sizes, companies can reduce the apparent quantities of calories, sugar, fats, or sodium displayed on the label, creating the illusion of a healthier product. For instance, a bag of chips might list a serving size as "15 chips," when in reality, most consumers are likely to consume half the bag or more in one sitting. Similarly, sugary beverages often define a single serving as a fraction of the container, leading to significant underestimation of caloric and sugar intake if the entire bottle is consumed (Kaur et al., 2017).

This practice is particularly misleading for products marketed as "snack-sized" or "individually portioned." Items such as granola bars or protein shakes often appear to be single servings but may list two or more servings per package. Consumers who fail to notice this detail might inadvertently double or triple their intake of less desirable nutrients, such as added sugars or saturated fats.

To counteract these deceptive tactics, it is important for consumers to critically assess serving sizes in relation to their actual consumption habits. Comparing the serving size to the amount typically eaten can provide a more accurate understanding of the nutritional content.

Daily Values (%DV): A Double-Edged Guide: Daily Values (%DV), another key element of nutritional labels, indicate how much a single serving of a food contributes to the recommended daily intake of specific nutrients. While these percentages are intended to help consumers gauge a product's nutritional value, they can be misleading without proper context. For example, the %DV is based on a 2,000-calorie diet, which may not align with the caloric needs of individuals with higher or lower energy requirements, such as athletes or those on weight-loss plans.

For beneficial nutrients like fiber, protein, vitamins, and minerals, higher %DV values are advantageous. A cereal providing 25% of the daily fiber requirement per serving, for instance, can be considered a nutrient-dense choice. Conversely, for nutrients like added sugars, sodium, and saturated fats, lower %DV values are preferable. However, many processed foods skew these metrics. A frozen dinner might boast 20% of the daily protein requirement but also contain 60% of the recommended sodium intake—a fact that could go unnoticed if consumers focus solely on the protein content.

Moreover, %DV values for added sugars have only recently been mandated in some countries, leaving gaps in consumer awareness. This can lead to an overreliance on unregulated claims like "low sugar," which may not align with actual dietary guidelines.

Empowering Consumer Choices: Interpreting serving sizes and %DV values is essential for making informed dietary choices, but the complexity of nutritional labels often deters even the most health-conscious individuals. Many consumers lack the knowledge or time to critically evaluate these details, leaving them vulnerable to marketing strategies designed to obscure unhealthy aspects of a product.

To address these challenges, education about ingredient transparency and label literacy is vital. Public health campaigns, clearer labeling standards, and tools like mobile apps that decode nutritional information can empower consumers to make better decisions. Teaching individuals how to interpret serving sizes in the context of real-life consumption and how to evaluate %DV values for their personal dietary needs can foster greater awareness and healthier habits.

By understanding the nuances behind serving sizes and %DV, consumers can cut through the confusion and take control of their nutritional choices, ultimately fostering a more transparent and health-oriented food environment.

Recognizing Greenwashing and "Health Halo" Marketing

In the competitive food industry, where health-conscious consumers represent a growing market segment, companies deploy sophisticated marketing strategies to position their products as healthy, sustainable, or environmentally friendly. Among these tactics, greenwashing and "health halo" marketing stand out as particularly deceptive practices that manipulate consumer perception and obscure the truth about a product's actual healthfulness or environmental impact.

Greenwashing: Misleading Environmental Claims

Greenwashing refers to the practice of using marketing to falsely portray a product or company as environmentally friendly. This tactic often involves vague or unsubstantiated claims that appeal to eco-conscious consumers without delivering meaningful environmental benefits. For example, terms like "sustainably sourced," "eco-friendly," or "biodegradable" are frequently used on product packaging, but without regulatory oversight or verification, these claims can be meaningless (Lyon & Maxwell, 2011).

Packaging plays a crucial role in greenwashing, with companies often using green color schemes, images of nature, or buzzwords to evoke an eco-friendly image. However, these visual cues can mislead consumers into believing that a product is sustainable when its production or disposal methods may still harm the environment. For instance, a product labeled as "biodegradable" may only degrade under specific industrial conditions, which are not accessible in most consumer settings.

To combat greenwashing, consumers should look for credible third-party certifications, such as the USDA Organic seal, Fair Trade certification, or Non-GMO Project Verified labels, which are governed by strict standards. Understanding the difference between these verified claims and marketing jargon is essential for making informed decisions.

The "Health Halo" Effect: Perceived Healthfulness

The "health halo" effect occurs when a single positive attribute of a product leads consumers to perceive it as healthier overall, often ignoring less desirable aspects of its nutritional profile. For example, products labeled as "low-fat," "gluten-free," or "organic" are often assumed to be healthier choices, even when they contain high levels of added sugar, sodium, or unhealthy fats (Chandon & Wansink, 2012).

Food companies capitalize on this effect by emphasizing trendy ingredients or health-related keywords on packaging. For instance, a cereal might highlight the inclusion of "superfoods" like quinoa or chia seeds in bold letters, while the fine print reveals high levels of sugar and artificial additives. Similarly, "plant-based" products may be marketed as inherently healthy, even when they are heavily processed and lack the nutritional balance of whole, plant-based foods.

The health halo effect can also lead to overconsumption. Consumers might feel less guilty about eating larger portions of a "low-fat" snack, not realizing that the calorie content per serving remains significant. This phenomenon underscores the importance of reading ingredient lists and nutritional panels rather than relying solely on front-of-package claims.

The Consequences of Misleading Marketing

Both greenwashing and health halo marketing contribute to consumer confusion, making it difficult to distinguish genuinely healthy or sustainable products from those that merely appear to be. These tactics can undermine public health initiatives and environmental efforts by shifting attention away from products that truly align with these goals.

For example, a consumer who chooses a "natural" granola bar over fresh fruit might inadvertently consume more sugar and calories, while assuming they are making a healthier choice. Similarly, purchasing a product with greenwashed claims might divert funds from companies that invest in legitimate sustainable practices, slowing progress toward environmental goals.

Moving Toward Transparency

Recognizing greenwashing and health halo marketing is the first step toward fostering greater transparency and accountability in the food industry. Consumers can protect themselves by learning to critically evaluate packaging, seeking out verified certifications, and prioritizing whole, minimally processed foods. Advocacy for clearer labeling standards and regulatory oversight can also help curb these deceptive practices, creating a marketplace where informed choices are easier to make.

By understanding these marketing tactics, consumers can resist manipulation, align their purchases with their values, and contribute to a food system that prioritizes health and sustainability over profit-driven deception.

Tips for Identifying Genuinely Healthier Options in a Sea of Processed Choices

Amid the overwhelming variety of processed food options, identifying genuinely healthier choices requires vigilance and knowledge. The following tips can help consumers make informed decisions:

- **Focus on Whole Ingredients**: Products with shorter ingredient lists and recognizable whole ingredients are typically healthier. For example, a granola bar made with oats, nuts, and honey is likely a better option than one with a long list of artificial additives.

- **Prioritize Nutritional Value**: Look for foods rich in fiber, protein, and essential vitamins and minerals. Avoid products with high amounts of added sugars, unhealthy fats, or sodium.

- **Beware of Deceptive Labels**: Terms like "organic" or "natural" do not automatically mean healthy. Organic cookies, for instance, may still contain significant amounts of sugar and fat. Instead, evaluate the overall composition of the product.

- **Learn to Recognize Additives**: Be cautious of artificial sweeteners, preservatives, and coloring agents. Ingredients like aspartame, sodium benzoate, and Red 40 can have health implications and are often unnecessary in whole-food-based diets.

- **Understand Certifications**: Familiarize yourself with credible certifications, such as Non-GMO Project Verified, USDA Organic, and Fair Trade. These labels offer a degree of assurance about a product's sourcing and quality.

- **Shop the Perimeter**: Grocery store layouts often place whole foods like fruits, vegetables, and fresh meats around the perimeter, while processed foods dominate the center aisles. Sticking to the perimeter can help prioritize healthier options.

- **Be Skeptical of Trends**: Trendy buzzwords like "keto-friendly" or "plant-based" are not guarantees of healthfulness. Read the ingredient list to determine whether the product aligns with your dietary needs and goals.

Navigating food labels and marketing myths is a critical skill in today's consumer landscape, where misleading claims and deceptive practices often obscure the truth about a product's health benefits or environmental impact. By learning to decode ingredient lists, decipher nutritional labels, recognize tactics like greenwashing and the "health halo" effect, and identify genuinely healthier options, consumers can make informed choices that align with their nutritional and ethical priorities. Education, vigilance, and critical thinking empower individuals to see beyond marketing façades, prioritize foods that support genuine healthfulness, and contribute to a more transparent food system. In doing so, we can foster better health outcomes for individuals and communities alike, making meaningful strides toward informed and sustainable dietary choices.

Chapter 14

Transforming the Modern Diet

Transforming the modern diet is a critical step toward improving health outcomes, fostering sustainability, and addressing the challenges posed by processed food consumption. While convenience and fast-paced lifestyles have led to the dominance of processed foods, the health and environmental consequences of this shift are undeniable. Reducing reliance on processed foods, embracing home-cooked meals, and leveraging innovations in food technology can pave the way for a more balanced and nutritious approach to eating. This chapter explores practical strategies for minimizing processed foods, the benefits of cooking from scratch, and the role of emerging technologies in creating healthier, more sustainable food options for the future.

Practical Advice for Reducing Processed Food Consumption

Reducing reliance on processed foods is a transformative step toward achieving a healthier diet and lifestyle. Processed foods, often laden with added sugars, unhealthy fats, and artificial additives, dominate modern diets due to their convenience and heavy marketing. However, taking intentional steps to minimize their role in daily meals can significantly improve overall health, lower the risk of chronic diseases, and contribute to better mental and physical well-being.

Start with Small, Sustainable Changes

Transitioning away from processed foods can feel daunting, especially for those accustomed to the convenience they offer. However, starting with small, sustainable changes can make the process more approachable and long-lasting. Begin by focusing on easy swaps: replace highly processed snacks like chips and cookies with whole-food alternatives such as nuts, fresh fruits, or plain yogurt. For example, substituting sugary breakfast cereals with oatmeal topped with fresh fruit can provide a healthier start to the day.

Gradual changes to staple ingredients can also have a profound impact. Swapping refined grains like white bread, pasta, and rice for their whole-grain counterparts, such as whole-grain bread, brown rice, or quinoa, preserves more fiber, vitamins, and minerals in your diet. Incremental shifts help individuals adapt without feeling deprived, making it easier to maintain these changes over time. As these habits solidify, they create a foundation for further dietary improvements (Slavin, 2013).

Read Labels and Avoid Hidden Additives

Understanding ingredient lists and nutritional labels is an essential skill for reducing processed food consumption. Many processed foods contain hidden additives designed to enhance flavor, extend shelf life, or improve texture, but these additives often have negative health implications. High-fructose corn syrup, artificial sweeteners, hydrogenated oils, and chemical flavorings are common culprits that should be minimized or avoided.

When reading labels, look for products with short ingredient lists composed of recognizable, whole-food ingredients. For example, a jar of peanut butter containing only "peanuts and salt" is preferable to one with added sugars, oils, and stabilizers. Similarly, avoid products with vague terms like "natural flavors," which can encompass a wide variety of unregulated additives (Campos et al., 2011).

Familiarizing yourself with nutritional information also helps. Check for high levels of sodium, added sugars, or unhealthy fats, as these are red flags indicating a highly processed product. By learning to decode these labels, consumers can make informed decisions and steer away from hidden additives.

Plan and Prioritize Home-Cooked Meals

The reliance on processed foods often stems from their convenience. To counteract this, planning and prioritizing home-cooked meals can make it easier to integrate healthier eating habits into busy schedules. Start by creating weekly meal plans that incorporate whole ingredients, such as fresh vegetables, lean proteins, and whole grains. A well-thought-out grocery list can minimize the temptation to purchase pre-packaged or processed foods.

Batch cooking is another practical strategy that ensures nutritious meals are always within reach. Dishes like soups, stews, casseroles, or grain bowls can be prepared in large quantities and stored for later use. Freezing portions for quick reheating during busy weekdays minimizes the need to rely on takeout or processed meals.

Meal planning also offers financial benefits by reducing food waste and eliminating impulse purchases. Over time, it fosters a sense of control over diet and nutrition, empowering individuals to maintain healthier eating patterns.

The Role of Meal Preparation and Cooking from Scratch

Cooking from scratch is a cornerstone of reducing processed food consumption, as it gives complete control over the ingredients and methods used. This approach not only enhances the nutritional quality of meals but also deepens an individual's connection to the food they consume. For instance, making homemade tomato sauce with fresh tomatoes, garlic, and herbs eliminates the added sugars and preservatives commonly found in store-bought versions.

Reclaiming the kitchen starts with simple recipes that do not require extensive skills or time. For example, roasting vegetables, preparing a basic stir-fry, or grilling lean proteins are quick and nutritious ways to create homemade meals. Over time, building confidence in the kitchen enables individuals to experiment with more complex recipes and diversify their diets.

Cooking from scratch also supports better portion control and customization to suit dietary needs or preferences. It can become a creative and rewarding process that fosters mindfulness about what goes into each meal, promoting healthier choices and reducing reliance on processed alternatives.

Reclaiming the Kitchen

Reclaiming the kitchen is a transformative step in fostering healthier eating habits and reducing reliance on processed foods. It begins with redefining home cooking as a creative and enjoyable activity rather than a burdensome chore. For many, cooking at home can feel intimidating, but starting with simple, accessible recipes can help build confidence and establish a sense of accomplishment.

Starting Simple

Simple recipes that require minimal ingredients and basic techniques are an excellent starting point for those new to cooking. Roasting vegetables with olive oil and herbs, preparing whole-grain salads with quinoa or barley, or creating a quick stir-fry with lean proteins and fresh vegetables are nutritious options that don't require extensive skills or time. These meals are customizable and allow individuals to experiment with flavors, textures, and ingredients, making the process both practical and rewarding.

Expanding Culinary Skills

As confidence grows, home cooks can expand their repertoire by experimenting with more complex dishes, exploring global cuisines, or learning advanced techniques such as baking bread or fermenting vegetables. This gradual progression not only diversifies the diet but also fosters a deeper appreciation for the culinary arts and the nutritional value of home-cooked meals. Over time, individuals can rely less on processed options, creating meals that are both healthier and more satisfying.

Utilizing Meal Prep Techniques

Meal preparation is a cornerstone of integrating home cooking into a busy lifestyle. By planning and preparing meals or components in advance, individuals can ensure they have access to nutritious, home-cooked options even on hectic days. Meal prep also streamlines the cooking process, making it more manageable and less time-consuming.

Batch Cooking

Batch cooking involves preparing large quantities of food that can be divided into individual portions and stored for later use. Dishes like soups, stews, casseroles, or cooked grains are ideal for this approach. By freezing individual servings, individuals can enjoy homemade meals without the daily time investment. This technique is particularly useful for families or professionals juggling packed schedules.

Prepping Ingredients in Advance

Prepping ingredients ahead of time further simplifies cooking. Tasks such as chopping vegetables, marinating proteins, or cooking staples like rice or beans can significantly reduce the time required to assemble meals during the week. For example, pre-cut vegetables can be used in salads, stir-fries, or omelets, while pre-cooked grains can serve as a base for grain bowls or side dishes.

Portion Control and Waste Reduction

Meal prep also promotes portion control, helping to prevent overeating and maintain a balanced diet. By dividing meals into pre-measured portions, individuals can avoid consuming excessive calories while still feeling satisfied. Additionally, planning meals reduces food waste by ensuring ingredients are used efficiently, which is both environmentally and financially beneficial (Pollan, 2008).

The Health and Social Benefits of Cooking at Home

Cooking at home offers numerous benefits that extend beyond individual health. It not only improves diet quality but also strengthens social bonds and cultural connections.

Improved Nutrition

Studies show that individuals who cook at home consume fewer calories, less sugar, and less fat compared to those who rely on restaurant or pre-packaged meals. Home cooking allows for greater control over ingredients, enabling the preparation of meals that are lower in sodium and unhealthy fats while incorporating more vegetables, whole grains, and lean proteins (Wolfson & Bleich, 2015).

Social Connection

Sharing home-cooked meals with family or friends fosters a sense of community and strengthens relationships. Mealtime can become a ritual for reconnecting, sharing stories, and passing down traditions. This cultural and familial exchange reinforces the value of food beyond its nutritional content, creating lasting memories and promoting mental well-being.

Empowerment and Autonomy

Cooking at home empowers individuals to take control of their diet and health. It builds confidence and autonomy, as individuals learn to create meals that suit their preferences, dietary restrictions, and lifestyle. This sense of agency is especially important in a food landscape dominated by convenience-driven processed options.

How Technology and Innovation Can Create Healthier Processed Foods

While reducing processed food consumption is ideal, advances in food technology have the potential to create healthier processed foods that align better with modern dietary needs. By reformulating ingredients and incorporating innovative techniques, food manufacturers can retain the convenience of processed foods while minimizing their negative health impacts.

Reformulating Processed Foods

Food manufacturers are actively reformulating their products to reduce or eliminate harmful ingredients like trans fats, added sugars, and excessive sodium. For instance, natural sweeteners such as stevia and monk fruit are increasingly being used to replace refined sugars, offering sweetness with fewer calories. Similarly, salt substitutes like potassium chloride are being incorporated to lower sodium content while maintaining flavor (Mozaffarian et al., 2018).

In addition to reducing harmful components, manufacturers are adding beneficial ones. Products fortified with fiber, omega-3 fatty acids, and probiotics are becoming more common, providing additional health benefits.

Functional and Personalized Foods

Functional foods, designed to provide specific health benefits beyond basic nutrition, are rapidly gaining traction as consumers increasingly prioritize health and wellness. These foods are intentionally formulated or enriched with bioactive compounds, vitamins, minerals, or other beneficial nutrients to address specific health concerns or enhance overall well-being. Meanwhile, the rise of personalized nutrition is revolutionizing how individuals approach their diets, using data-driven insights to create foods tailored to individual needs, preferences, and even genetic profiles. Together, these innovations have the potential to redefine processed foods, turning them into powerful tools for promoting health rather than contributors to dietary imbalances.

Functional Foods: Addressing Specific Health Needs

Functional foods are designed to bridge the gap between nutrition and medicine by offering targeted health benefits. Examples of functional foods include:

- **Probiotic-Rich Yogurts**: Containing live active cultures, these products promote gut health by supporting a balanced microbiome. Improved gut health has been linked to better digestion, enhanced immunity, and even mental health benefits through the gut-brain axis.

- **Fortified Cereals and Plant-Based Milks**: Many cereals are enriched with iron and folic acid to prevent anemia, while plant-based milks are fortified with calcium and vitamin D to support bone health, making them suitable dairy alternatives for individuals with lactose intolerance or dietary restrictions.

- **Omega-3-Enriched Eggs and Dairy Products**: These foods contribute to cardiovascular health by delivering heart-healthy fats that reduce inflammation and lower cholesterol levels.

- **Functional Beverages**: Drinks enhanced with antioxidants, adaptogens, or electrolytes cater to a range of needs, from improving focus and reducing stress to replenishing hydration after physical activity.

The popularity of functional foods stems from their ability to target specific health concerns while seamlessly integrating into daily routines. For example, a consumer seeking improved gut health might choose a yogurt with added probiotics over traditional varieties, or someone managing cholesterol may opt for margarine enriched with plant sterols.

The Evolution of Personalized Nutrition

Personalized nutrition represents the next frontier in food innovation, leveraging technology and data to create tailored dietary solutions. This approach moves beyond one-size-fits-all recommendations, instead focusing on individual differences in metabolism, health conditions, and lifestyle preferences.

Advances in genetic testing have played a pivotal role in the evolution of personalized nutrition. Companies like 23andMe and DNAfit offer insights into how genetic variations influence dietary needs, such as susceptibility to lactose intolerance, gluten sensitivity, or the efficiency of metabolizing caffeine. With this information, consumers can make more informed choices, such as avoiding certain foods or incorporating specific nutrients to optimize their health.

Wearable devices and nutrition apps further enhance personalization by tracking real-time data on activity levels, calorie expenditure, and macronutrient intake. For example, a fitness enthusiast might use these tools to ensure they are consuming sufficient protein to support muscle recovery, while an individual managing diabetes can receive guidance on maintaining stable blood sugar levels.

In response to this demand for personalization, food companies are developing products that cater to specific dietary goals. For instance, high-protein snacks for athletes, low-glycemic foods for individuals with diabetes, and allergen-free products for those with food sensitivities are becoming increasingly available. Subscription-based services offering customized meal kits tailored to dietary preferences and health goals are also growing in popularity, providing convenience without sacrificing individuality.

Transforming Processed Foods into Health Promoters

The convergence of functional and personalized nutrition has the potential to transform processed foods into agents of health promotion. Rather than contributing to dietary imbalances, these foods can help address nutrient deficiencies, manage chronic conditions, and support overall well-being.

For example, a functional protein bar enriched with prebiotics and tailored to an individual's activity level could serve as a post-workout snack, supporting gut health and muscle recovery simultaneously. Similarly, a meal replacement shake designed for an older adult with specific nutrient needs, such as increased calcium and vitamin D, could support bone density and overall vitality.

The potential for personalized functional foods extends to population-level health improvements. By addressing specific needs in a targeted and efficient manner, these innovations could help combat global challenges such as obesity, malnutrition, and diet-related chronic diseases. For instance, fortified staple foods like rice or flour can address micronutrient deficiencies in vulnerable populations, while personalized meal solutions can cater to the unique needs of individuals managing conditions like cardiovascular disease or autoimmune disorders.

The Future of Functional and Personalized Foods

As consumer demand for health-focused food options continues to grow, the functional and personalized food markets are expected to expand rapidly. Advances in technology, including artificial intelligence and machine learning, are likely to drive further innovation in this space. AI-powered tools can analyze large datasets on diet, genetics, and health outcomes to recommend optimal nutrient profiles or develop new functional food formulations.

Collaboration between food scientists, healthcare providers, and technology companies will be essential in ensuring these innovations are both effective and accessible. With the right balance of science, technology, and consumer engagement, functional and personalized foods can pave the way for a future where processed foods actively contribute to improved health and well-being on an individual and global scale.

Healthier processed foods are increasingly adopting sustainable practices to address the dual challenges of feeding a growing population and mitigating environmental damage. With food production accounting for a significant share of global greenhouse gas emissions, land use, and water consumption, innovative approaches are reshaping how food is produced, processed, and delivered. These practices aim to create a more resilient and eco-friendly food system while maintaining nutritional quality and convenience.

The Rise of Plant-Based Proteins

Plant-based proteins have emerged as a cornerstone of sustainable food innovation. Products derived from sources such as soy, peas, lentils, and chickpeas offer nutrient-dense alternatives to animal-based proteins with a significantly smaller environmental footprint. Compared to traditional livestock farming, plant-based protein production requires less land, water, and energy, while emitting fewer greenhouse gases. Popular brands like Beyond Meat and Impossible Foods have revolutionized this market by creating products that closely mimic the taste, texture, and appearance of meat, appealing to both vegetarians and meat-eaters seeking to reduce their environmental impact.

The appeal of plant-based proteins extends beyond individual health and sustainability. These products often incorporate fortifications such as added vitamins, minerals, and fiber, making them nutritionally competitive with traditional protein sources. Additionally, the scalability of plant-based protein production positions it as a viable solution for feeding a growing global population without exacerbating environmental degradation.

Lab-Grown Meats: The Future of Protein

Lab-grown meats, also known as cultured or cell-based meats, represent a groundbreaking innovation in sustainable food production. By cultivating animal cells in controlled environments, lab-grown meat eliminates the need for raising and slaughtering livestock. This technology drastically reduces the environmental footprint of meat production, including land use, water consumption, and methane emissions associated with livestock farming.

In addition to environmental benefits, lab-grown meat addresses ethical concerns related to animal welfare, offering consumers a guilt-free alternative to conventionally sourced meat. While still in its early stages, this technology is rapidly advancing, with several companies bringing prototypes to market. As production scales and costs decrease, lab-grown meat has the potential to become a mainstream alternative, contributing to a more sustainable and ethical food system.

Exploring Alternative Protein Sources

Beyond plant-based and lab-grown options, alternative proteins derived from unconventional sources such as algae and insects are gaining traction. Algae, particularly microalgae like spirulina and chlorella, are rich in protein, vitamins, and essential fatty acids, while being highly efficient to cultivate. Algae can grow in diverse environments, including saline or nutrient-poor waters, making it a sustainable crop that doesn't compete with traditional agriculture for arable land.

Insect protein, made from species like crickets or mealworms, offers another promising alternative. Insects are highly efficient at converting feed into protein, require minimal space and water, and emit negligible greenhouse gases. Products such as cricket flour are already being incorporated into energy bars, snacks, and even baked goods, appealing to environmentally conscious consumers and adventurous eaters alike.

Reducing Packaging Waste

Sustainability in processed foods extends beyond ingredient choices to include packaging innovations. Single-use plastics, a major contributor to environmental pollution, are being replaced with biodegradable, compostable, or recyclable materials. Edible packaging made from seaweed or starch-based materials is another emerging solution, eliminating waste altogether by allowing the packaging to be consumed along with the food.

Minimalist packaging, which uses fewer materials or eliminates unnecessary layers, is also gaining popularity. Companies are leveraging innovative designs to reduce waste while maintaining product safety and shelf life. These efforts not only reduce environmental harm but also resonate with consumers who value eco-friendly practices.

Impact on Health and the Environment

Sustainable innovations in processed foods address health and environmental concerns simultaneously. By integrating nutrient-dense, eco-friendly ingredients and reducing waste, these practices support individual well-being while mitigating the global challenges of climate change, resource scarcity, and biodiversity loss. The alignment of health and sustainability in food production is not only a necessity but also a powerful opportunity to drive positive change on a global scale.

Fortification and Functional Foods

One promising area of innovation in food technology is the development of functional foods—processed products that are intentionally enhanced with beneficial nutrients to address specific health needs. These foods serve dual purposes, providing sustenance while also delivering added health benefits, making them valuable tools in addressing nutritional gaps in modern diets. For example, cereals fortified with iron and vitamin D can combat deficiencies commonly found in certain populations, particularly among children and individuals in regions with limited sun exposure. Fortification of these nutrients can significantly reduce the risk of anemia and rickets, conditions that remain prevalent in many parts of the world.

Similarly, plant-based milk alternatives, such as almond, soy, and oat milk, are enriched with calcium and vitamin B12 to mimic the nutritional profile of dairy milk. These enhancements make plant-based options viable for those with lactose intolerance or dietary restrictions while ensuring they do not miss out on essential nutrients. Functional foods can also target specific health conditions, such as probiotic-rich yogurts that support gut health or omega-3-enriched spreads that promote cardiovascular health.

Beyond individual health benefits, these fortified products can play a critical role in public health strategies. In populations with limited access to fresh, nutrient-dense whole foods, fortified products offer a practical solution for preventing deficiencies and improving overall nutrition. For example, iodized salt has been a global success story, dramatically reducing iodine deficiency disorders. Similarly, vitamin A-enriched staples like rice or cooking oil are being utilized to combat blindness and immune deficiencies in vulnerable populations. As food science continues to advance, the scope and effectiveness of functional foods will likely expand, addressing an even broader range of dietary needs and health challenges.

Leveraging Technology for Personalization

The emergence of personalized nutrition is revolutionizing the way we approach food and health. Advances in technology, including apps, wearable devices, and genetic testing, are enabling tailored dietary recommendations based on an individual's specific health needs, preferences, and even genetic makeup. This level of personalization allows consumers to make more informed food choices that align with their unique goals, whether they aim to lose weight, manage a chronic condition, or optimize athletic performance.

Wearable devices, such as fitness trackers and smartwatches, can monitor daily activities, sleep patterns, and dietary habits. By syncing this data with nutrition apps, users receive customized recommendations that account for their caloric expenditure, nutrient intake, and personal health metrics. For instance, a person looking to manage diabetes might receive suggestions for low-glycemic index foods, while an athlete training for a marathon might be directed toward high-protein snacks and carbohydrate-rich meals to fuel their performance.

Genetic testing takes personalization a step further by identifying how an individual's genetic profile influences their nutritional needs. Companies like 23andMe and DNAfit offer insights into how genes affect metabolism, food sensitivities, and nutrient absorption. This information allows for the creation of diets tailored to genetic predispositions, such as lactose intolerance or a tendency toward elevated cholesterol levels.

In the food industry, companies are responding to the demand for personalized nutrition by developing products that cater to specific dietary goals. Protein-enriched snacks for athletes, fiber-rich foods for digestive health, and plant-based alternatives for eco-conscious consumers are just a few examples. Personalized meal kits, which combine convenience with customized nutrition, are also growing in popularity, enabling consumers to prepare meals that meet their unique dietary needs without the guesswork.

This integration of technology into food choices not only empowers consumers but also has the potential to improve public health by making precision nutrition more accessible and actionable. As these tools continue to evolve, they are likely to play a pivotal role in bridging the gap between processed foods and individualized health.

Sustainability in Processed Foods

The push for healthier processed foods is increasingly intertwined with the need for sustainable food production practices. As the global population grows and environmental concerns intensify, the food industry is exploring innovative ways to reduce its ecological footprint while maintaining nutritional quality and convenience.

Plant-based proteins have emerged as a cornerstone of sustainable food innovation. Products derived from soy, peas, and chickpeas are not only rich in protein but also require significantly fewer resources to produce compared to animal-based proteins. Companies like Beyond Meat and Impossible Foods have popularized meat alternatives that mimic the texture and flavor of traditional meats, offering environmentally conscious consumers satisfying options that align with their values. These alternatives help reduce greenhouse gas emissions, land use, and water consumption associated with conventional livestock farming.

In addition to plant-based proteins, alternative protein sources derived from insects or algae are gaining attention. Crickets, for example, are a sustainable and nutrient-dense source of protein, requiring minimal resources to farm. Algae, which can be grown in diverse environments and has a high yield per unit area, offers essential nutrients such as omega-3 fatty acids and antioxidants. Incorporating these unconventional ingredients into processed foods could revolutionize the industry by creating products that are both nutritious and environmentally friendly.

Lab-grown meats, also known as cultured meats, represent another groundbreaking innovation in sustainable food production. By cultivating animal cells in a controlled environment, lab-grown meat eliminates the need for traditional farming practices, reducing the ethical and environmental issues associated with livestock agriculture. While still in its early stages, this technology has the potential to provide high-quality protein with a fraction of the environmental impact of conventional methods (Post et al., 2020).

Sustainability in processed foods also extends to packaging. Efforts to reduce single-use plastics and adopt biodegradable or compostable materials are addressing the significant waste generated by the food industry. Companies are exploring edible packaging, such as seaweed-based wraps, and innovations in minimalistic packaging to reduce environmental harm.

By integrating sustainability into processed food production, the industry is addressing the dual challenges of feeding a growing population and protecting the planet. These innovations not only support environmental goals but also align with the values of an increasingly eco-conscious consumer base, paving the way for a more sustainable future.

Transforming the modern diet requires a combination of individual action, societal change, and technological innovation. By reducing processed food consumption, embracing home cooking, and supporting advancements in food technology, we can create a healthier, more sustainable food culture. Whether through small, practical steps in the kitchen or large-scale changes in food production, each effort contributes to reshaping the way we eat and improving health outcomes for future generations.

Chapter 15

Point of No Return
Why We Can't Fully Go Back

The modern food system has undergone a profound transformation over the last century, shaped by industrialization, globalization, and cultural shifts. This evolution has enabled humanity to feed billions, yet it has also entrenched a dependence on processed foods, industrial agriculture, and global supply chains. While many recognize the health and environmental consequences of these systems, returning to pre-industrial methods is neither practical nor feasible. The economic reliance on the processed food industry, the deep integration of convenience into daily life, and the irreversible changes to ecosystems and infrastructure highlight why the world has reached a point of no return. Understanding these realities is essential to finding sustainable ways forward within the constraints of our modern food system.

The Industrialization of Food: How Modern Agriculture and Processing Have Become Integral to Feeding Billions

The industrialization of food production represents one of humanity's most significant technological achievements, revolutionizing how societies grow, process, and consume food to meet the needs of a rapidly expanding global population. Today, more than 8 billion people rely on food systems that combine modern agricultural practices and advanced processing technologies to ensure a steady supply of affordable and accessible food. High-yield crop varieties, monoculture farming, and the extensive use of fertilizers and pesticides have driven unprecedented levels of agricultural productivity, transforming barren lands into breadbaskets capable of sustaining entire nations. These advancements, alongside innovations in food storage, preservation, and transportation, have made it possible to produce and distribute vast quantities of food with remarkable efficiency (Foley et al., 2011).

Food processing technologies, including canning, freezing, and packaging, have further enhanced the industrialized food system's capacity to reduce waste and extend shelf life. These innovations have enabled global trade networks to deliver products to the most remote corners of the world, addressing hunger and malnutrition in previously underserved regions. Convenience and affordability have made processed foods an integral part of modern life, shaping diets and lifestyles across all socioeconomic strata.

However, the success of industrialized food systems has come with significant costs. The emphasis on maximizing yields and efficiency has led to environmental degradation on a massive scale. Monoculture farming, for instance, depletes soil nutrients and reduces biodiversity, leaving ecosystems vulnerable to pests and diseases. The overuse of synthetic fertilizers and pesticides has polluted water sources, while the reliance on fossil fuels for machinery, transportation, and processing has made food production a major contributor to greenhouse gas emissions. According to the United Nations Food and Agriculture Organization (FAO), agriculture and land use changes account for nearly a quarter of global emissions, further exacerbating climate change.

In addition to environmental concerns, industrialized food systems are increasingly criticized for their impact on public health and social equity. The widespread availability of calorie-dense, nutrient-poor processed foods has contributed to rising rates of obesity, diabetes, and other diet-related diseases, even as malnutrition persists in many regions. Moreover, the centralization of food production has marginalized smallholder farmers and disrupted traditional agricultural practices, eroding local food sovereignty.

Despite these challenges, the scale and efficiency of industrialized food systems have made them indispensable. Reverting to pre-industrial methods, such as small-scale organic farming and local food production, would face significant obstacles. These methods, while more sustainable in theory, lack the capacity to feed billions of people consistently and affordably. A widespread return to traditional practices could result in dramatic food shortages, soaring prices, and increased vulnerability to natural disasters and climate variability.

This tension between sustainability and necessity highlights the complexity of the global food system. While there is a growing need to address the environmental and social consequences of industrialized food production, dismantling the existing system is not a viable option. Instead, efforts must focus on reforming and innovating within the current framework, exploring solutions such as regenerative agriculture, sustainable intensification, and technological advancements that balance productivity with ecological stewardship. By acknowledging both the achievements and the shortcomings of industrialized food systems, societies can work toward a more equitable and sustainable future for global food production.

Economic Realities: The Dependence of Economies and Jobs on the Processed Food Industry

The processed food industry is not just a vital component of modern economies—it is a cornerstone, shaping the global economic landscape through its immense contribution to GDP, job creation, and interconnected industries. Globally, this sector generates billions of dollars annually, supporting millions of workers across production, distribution, retail, and associated services. From sprawling manufacturing plants to small businesses supplying raw materials, the economic infrastructure built around processed foods is extensive and deeply ingrained. For example, in the United States, the food and beverage industry alone employs over 1.7 million people and contributes more than $1 trillion annually to the national economy (IBISWorld, 2023). This expansive industry encompasses everything from agriculture and logistics to advertising and retail, creating a web of economic activity that fuels local and national economies.

In developing nations, the processed food industry holds an equally critical role, driving industrialization and fostering economic growth. Urbanization and rising middle-class incomes in countries like India, Brazil, and China have accelerated the demand for packaged foods, integrating these economies into global supply chains. In India, for instance, the food processing sector is one of the largest in the country, employing millions and accounting for a significant share of manufacturing GDP. The growth of this sector has spurred ancillary industries, such as packaging, logistics, and retail, amplifying its economic impact. Similarly, in Brazil, the expansion of processed food production has created jobs in agriculture, manufacturing, and exports, making it a key driver of economic progress.

The interconnected nature of the processed food industry also highlights its critical role in stabilizing food supply chains. Industrial food production ensures a consistent and reliable supply of affordable food products, meeting the demands of urban populations and mitigating food insecurity in regions with limited agricultural output. Packaged and processed foods are particularly important in disaster-prone or resource-constrained areas, where their long shelf life and ease of transport make them indispensable for food distribution efforts.

However, this economic dependency also presents challenges. Disrupting or dismantling the processed food industry would have far-reaching consequences, including widespread job losses, economic instability, and the collapse of supply chains that many rely on for essential goods. The ripple effects would be felt across multiple sectors, from transportation to retail, exacerbating economic inequalities and creating significant social challenges.

Moreover, the global interconnectedness of the processed food industry means that its disruption could destabilize international trade and economic relationships. Developing countries that depend on food exports for economic growth could face severe setbacks, while consumer economies in developed nations would experience shortages and price surges. These economic realities make a complete return to traditional food systems not only impractical but potentially catastrophic.

Instead, the focus must shift toward reform and innovation within the existing framework. Strategies such as improving sustainability practices, reducing the environmental footprint of food production, and addressing public health concerns through product reformulation can help balance the economic benefits of the processed food industry with its broader societal and environmental impacts. By embracing technological advancements and regulatory oversight, the industry can evolve to meet modern challenges while preserving its role as a cornerstone of global economic stability.

Cultural Shifts: Convenience and Time-Saving Meals as Staples of Modern Lifestyles

The rise of processed foods is deeply intertwined with cultural shifts that prioritize convenience and efficiency, reflecting the fast-paced realities of modern life. As individuals and families juggle work, education, childcare, and other responsibilities, time-saving meals and snacks have become indispensable. Ready-to-eat meals, frozen dinners, and pre-packaged snacks cater to these demands, offering practical solutions for those who lack the time or energy to cook from scratch. For many, these foods are not just a matter of convenience but a necessity in managing the complexities of daily life.

The marketing of convenience foods has further entrenched their role in contemporary culture. Advertising campaigns often depict processed foods as essential for busy lifestyles, framing them as symbols of modernity and aspiration. Slogans emphasizing speed, ease, and enjoyment position these products as solutions to the challenges of modern living, reinforcing their appeal. This has led to the gradual erosion of cooking traditions that once defined cultural identities. Recipes and techniques passed down through generations have been supplanted by store-bought alternatives, shifting the focus from culinary heritage to efficiency (Pollan, 2008).

This reliance on processed foods has reshaped the way food is perceived and consumed, making it a central component of urban and suburban life. In densely populated areas, where time and resources are often constrained, traditional meal preparation methods have become less viable. The accessibility and affordability of processed options further reinforce their dominance, as they are often easier to obtain than fresh ingredients or home-cooked meals.

Efforts to encourage a return to traditional cooking must address the deeply ingrained cultural and practical factors driving this reliance on convenience foods. Educational campaigns promoting the benefits of home cooking, improved access to fresh ingredients, and time-saving techniques for meal preparation could help bridge the gap. However, these initiatives must be sensitive to the realities of modern life, recognizing that convenience foods have become more than a dietary choice—they are a cultural cornerstone that reflects the priorities and pressures of the contemporary world.

Globalization of Diets: How Processed Foods Have Reshaped Eating Habits Worldwide

The globalization of food systems has significantly reshaped diets across the globe, driven largely by the proliferation of processed foods. Global brands like Nestlé, Coca-Cola, and McDonald's have introduced uniform food products that transcend cultural and geographic boundaries, embedding themselves in markets worldwide. As these brands expand their reach, local cuisines are increasingly replaced or supplemented by processed and packaged alternatives. Even in countries with rich culinary traditions, such as Japan, the influence of global food systems has become evident. Japanese diets, traditionally centered around fresh fish, rice, and vegetables, are increasingly incorporating Western-style processed foods, including fast food, sugary beverages, and pre-packaged snacks (Popkin et al., 2020).

This phenomenon, often referred to as the "nutrition transition," has brought both benefits and challenges. On the one hand, processed foods have improved access to affordable calories in developing regions, alleviating hunger and malnutrition in some areas. On the other hand, they have contributed to a surge in diet-related health issues, such as obesity, diabetes, and cardiovascular diseases. For instance, the introduction of sugary beverages and ultra-processed snacks has led to dramatic increases in sugar consumption and caloric intake in countries where these products were previously unavailable, amplifying public health concerns.

Globalization has also led to a homogenization of diets, with traditional food practices and recipes increasingly overshadowed by mass-produced, standardized alternatives. This dietary convergence has reduced the diversity of ingredients and preparation methods that once characterized regional cuisines, eroding cultural heritage in the process. In Japan, for example, younger generations are more likely to consume Western-style fast food than traditional meals like miso soup and grilled fish, reflecting a shift driven by convenience, urbanization, and global marketing efforts.

The dominance of multinational corporations further solidifies this trend, as they wield significant influence over global food supply chains. Their ability to standardize production and distribution has made processed foods ubiquitous and deeply entrenched in daily life.

Reversing this shift is challenging, as the economic and cultural systems supporting globalized diets are robust and well-established.

Efforts to preserve traditional diets and combat the health effects of processed foods must navigate the complexities of globalization. Initiatives that promote local food systems, educate consumers about nutrition, and adapt traditional recipes for modern lifestyles could help counteract the homogenization of diets. However, these efforts face significant obstacles, as the global food industry continues to prioritize convenience and profit over cultural and dietary diversity.

Irreversible Changes: How Land Use, Biodiversity Loss, and Industrial Infrastructure Make Reverting Impractical

The industrialization of food systems has brought about irreversible changes to land use, biodiversity, and infrastructure, further complicating efforts to return to traditional food production methods. Monoculture farming, for instance, has replaced diverse ecosystems with single-crop landscapes, leading to habitat destruction and a significant decline in biodiversity (Tilman et al., 2001). These changes have created dependencies on high-yield crops like wheat, rice, and corn, which now dominate global agriculture.

Additionally, the infrastructure supporting modern food systems—from sprawling factories and distribution networks to refrigeration and retail—has been optimized for the production and delivery of processed foods. Transitioning back to decentralized, traditional food systems would require massive overhauls of these networks, a financially and logistically daunting task.

Finally, the widespread use of synthetic fertilizers and pesticides has altered soil chemistry and agricultural practices, creating a reliance on industrial inputs that cannot be quickly undone. The scale and permanence of these changes underscore the challenges of reverting to pre-industrial food systems, emphasizing the need for sustainable innovations within the existing framework.

The modern food system, with its reliance on industrialized agriculture and processed foods, has reached a point of no return. Economic dependencies, cultural shifts, globalized diets, and irreversible environmental changes have made a complete reversion to traditional methods impractical. However, acknowledging these realities does not mean abandoning efforts for reform. Instead, it highlights the importance of advancing sustainable practices, improving health outcomes, and innovating within the current system to address the pressing challenges of feeding a growing global population.

Chapter 16

Reclaiming Control
A New Way Forward

Reclaiming control over our modern food system is both a challenge and an opportunity, requiring a shift in how we define progress, innovate, and engage as consumers. For decades, the global food system has prioritized efficiency, convenience, and profit at the expense of health, sustainability, and cultural diversity. Now, as the consequences of these priorities become increasingly evident—from rising obesity rates to environmental degradation—it is clear that a new path forward is needed. This chapter explores the pillars of a reimagined food system: redefining progress, leveraging innovation, empowering consumers through food literacy, and implementing hybrid solutions that blend the benefits of processing with the nutrition of whole foods. Additionally, it examines the role of policy and regulation in creating a framework for a healthier and more equitable food future. Together, these strategies provide a roadmap for reclaiming control and creating a balanced, resilient food system for generations to come.

Redefining Progress: What Would "Progress" Look Like in a Balanced Food System?

Progress in the context of food systems must go beyond the narrow lens of productivity and profitability to incorporate health, sustainability, and cultural preservation. A balanced food system would prioritize equitable access to nutritious, affordable food while minimizing environmental impacts and preserving cultural traditions. This redefinition of progress requires a holistic approach that integrates public health goals, environmental stewardship, and economic equity.

In such a system, food production would adopt regenerative agricultural practices that restore soil health, enhance biodiversity, and reduce greenhouse gas emissions. Food distribution would focus on reducing waste and ensuring that underserved communities have access to fresh, nutrient-dense options. Technological innovations would align with these values, creating processes that are both efficient and ecologically responsible. Progress in this model is not just about feeding more people but about feeding them better, ensuring that every bite supports personal health and planetary well-being (Foley et al., 2011).

The Role of Innovation: Breakthroughs in Healthier Processing, Lab-Grown Ingredients, and Biodegradable Packaging

Innovation lies at the heart of transforming the modern food system into one that is both healthier for individuals and more sustainable for the planet. Advances in food science and technology offer promising pathways for refining processing methods, developing alternative ingredients, and rethinking packaging to address environmental concerns. Together, these breakthroughs provide an opportunity to reshape how food is produced, consumed, and disposed of, paving the way for a more equitable and resilient food system.

Healthier Processing Techniques

Modern food processing methods are evolving to address the health concerns associated with traditional processing techniques. Many older methods, such as deep frying, canning, and over-reliance on chemical preservatives, can degrade nutrients and introduce harmful additives into packaged foods. In response, newer technologies are focusing on retaining natural nutritional value while enhancing safety and shelf life.

High-Pressure Processing (HPP) is one such innovation. This technique uses high pressure instead of heat to eliminate harmful bacteria and pathogens, preserving the product's natural flavors, vitamins, and minerals. Unlike traditional pasteurization methods, which can strip away nutrients, HPP ensures that foods like juices, deli meats, and dairy retain their original quality while extending shelf life (Knorr et al., 2011).

Vacuum Frying is another breakthrough, particularly for snack foods. By frying foods at lower temperatures under vacuum conditions, this method reduces the absorption of oil, resulting in snacks with significantly lower fat content. This technique is especially useful for preserving the natural colors, flavors, and nutrients of fruits and vegetables.

Other innovations include freeze-drying and infrared heating, both of which maintain the integrity of sensitive nutrients while enhancing shelf life. These techniques allow for the creation of convenient, packaged foods that are far closer to whole foods in terms of nutritional quality.

Lab-Grown Ingredients

Lab-grown ingredients represent a revolutionary shift in how food is produced, offering solutions to both ethical and environmental challenges. Among the most prominent innovations in this area are cultured meats and precision fermentation technologies.

Cultured Meat, also known as lab-grown or cell-based meat, is produced by cultivating animal cells in a controlled environment, eliminating the need for traditional livestock farming. This technology addresses ethical concerns related to animal welfare and significantly reduces the environmental footprint of meat production. Studies suggest that cultured meat could cut greenhouse gas emissions by up to 96% compared to conventional beef farming, while requiring far less land and water (Post et al., 2020).

Plant-Based Proteins enhanced through precision fermentation are another groundbreaking development. Precision fermentation involves programming microorganisms like yeast or fungi to produce specific proteins, such as casein or whey, which mimic the properties of animal-derived ingredients. These proteins can be used to create plant-based alternatives for cheese, milk, and meat, offering consumers the taste and texture of traditional animal products with a fraction of the environmental impact.

These innovations not only provide sustainable alternatives to resource-intensive foods but also allow for customization. For instance, lab-grown meats and proteins can be tailored to include higher levels of beneficial nutrients, such as omega-3 fatty acids, or to exclude allergens, creating healthier and more inclusive options for consumers.

Biodegradable Packaging

As environmental concerns surrounding plastic pollution escalate, the food industry is turning to innovative packaging solutions to reduce its ecological footprint. Single-use plastics, which dominate the processed food market, contribute significantly to global waste and marine pollution. Biodegradable and compostable materials are being developed to address these issues, offering sustainable alternatives that align with circular economy principles.

Seaweed-Based Packaging has emerged as a promising solution, with companies creating flexible films that are not only biodegradable but also edible. This dual-purpose innovation eliminates waste entirely, making it ideal for single-serve products such as condiment packets or snack wraps.

Cornstarch and Sugarcane Packaging are also gaining traction as replacements for traditional plastics. These materials break down naturally in composting conditions, reducing the long-term environmental impact. For example, sugarcane-based containers are increasingly being used for takeout packaging, providing a sturdy yet eco-friendly alternative to Styrofoam.

Edible Films made from rice or potato starch add another layer of sustainability by integrating directly into the food supply chain. These films can be consumed along with the food they package, further minimizing waste. Such innovations not only reduce the environmental impact of food packaging but also resonate with eco-conscious consumers who value sustainability in their purchasing decisions (Rujnić-Sokele & Pilipović, 2017).

Biodegradable packaging is more than a technological advancement; it represents a shift in how the food industry addresses its responsibility toward environmental stewardship. By incorporating these materials into mainstream production, companies can significantly reduce waste while meeting the demands of sustainability-focused consumers.

Empowering Consumers: Building a Culture of Food Literacy

Consumers hold immense power to influence the food industry, but leveraging this power requires education and awareness. Building a culture of food literacy is essential for enabling individuals to make informed choices about their diets, demand better practices from producers, and advocate for systemic change.

Food Literacy Defined

Food literacy encompasses a range of skills and knowledge, including the ability to interpret nutritional labels, understand ingredient lists, recognize marketing tactics, and evaluate the environmental impact of food choices. It also involves practical skills, such as cooking and meal planning, which empower individuals to rely less on processed foods and more on whole, minimally processed alternatives.

The Role of Education

Public education campaigns play a crucial role in fostering food literacy. Initiatives that teach label literacy enable consumers to identify harmful additives, excessive sugars, and unhealthy fats. For example, a campaign that explains how to decipher front-of-package claims like "low-fat" or "all-natural" can prevent consumers from falling prey to misleading marketing tactics.

Community Engagement

Local programs, such as urban gardening projects and farmers' markets, reconnect communities with the sources of their food. These initiatives not only provide fresh, affordable produce but also create opportunities for individuals to learn about sustainable practices and the value of local food systems.

Long-Term Impact

Empowered consumers drive demand for healthier, more sustainable food options, incentivizing companies to prioritize transparency and innovation. As food literacy spreads, it has the potential to reshape the food industry, creating a marketplace that aligns more closely with the values of health, sustainability, and equity.

Educational Campaigns

Public education campaigns are essential for fostering food literacy and empowering consumers to make informed decisions in a complex food system. These initiatives aim to bridge the knowledge gap around nutrition, food sourcing, and environmental sustainability, equipping individuals with the tools they need to navigate modern dietary challenges.

Label Literacy and Nutritional Awareness

A cornerstone of food literacy is understanding how to interpret ingredient lists and nutritional labels. Educational programs that teach label literacy empower consumers to identify and avoid harmful additives, excessive sugars, and unhealthy fats. For instance, many processed foods disguise high sugar content under names like "dextrose" or "maltose," which can confuse consumers. By learning to recognize these terms and assess portion sizes, individuals can make healthier choices and avoid falling prey to deceptive marketing practices.

Interactive workshops, webinars, and school-based programs are effective platforms for promoting label literacy. Schools, in particular, play a crucial role in teaching children how to critically evaluate food options, setting the foundation for lifelong healthy eating habits. Public campaigns, such as "Eat Smart" or "Read Before You Eat," can extend these lessons to broader audiences through social media, public service announcements, and community events.

Cooking Skills and Practical Nutrition

Cooking skills are another critical component of food literacy. Many people rely on processed foods due to a lack of confidence or experience in the kitchen. Educational campaigns that focus on practical cooking skills can demystify meal preparation, showing individuals how to create wholesome meals quickly and affordably.

For example, cooking classes can teach simple techniques like roasting vegetables, preparing soups, or making homemade sauces. Online resources, including video tutorials and recipe platforms, make it easier than ever for people to access step-by-step guidance. Programs tailored to specific demographics, such as busy parents or college students, can address unique challenges and help participants integrate healthier habits into their daily routines (Pollan, 2008).

Highlighting Environmental Impact

Educational campaigns that emphasize the environmental consequences of food choices can also drive meaningful change. By illustrating the connection between dietary habits and issues like deforestation, water pollution, and carbon emissions, these initiatives encourage consumers to prioritize sustainable options. For instance, campaigns might highlight the benefits of reducing meat consumption or choosing foods with minimal packaging, fostering a more eco-conscious approach to eating.

Community Engagement

Community-driven programs take education a step further by creating opportunities for individuals to actively participate in food production and sourcing. These initiatives reconnect people with the origins of their food, promoting awareness of sustainability and empowering them to take control of their dietary choices.

Urban Gardens and Local Food Systems

Urban gardening projects bring fresh produce to city environments, addressing issues like food deserts and limited access to nutritious options. Community gardens allow residents to grow their own fruits and vegetables, fostering a sense of ownership and pride in their food supply. These gardens also provide educational opportunities, teaching participants about organic farming, composting, and the importance of biodiversity in food production.

Farmers' markets play a similar role by connecting consumers directly with local producers. By supporting small-scale farmers, these markets promote economic sustainability while providing fresh, affordable food. Shoppers also gain a deeper appreciation for seasonal and regional foods, encouraging them to prioritize fresh ingredients over processed alternatives.

Social Cohesion and Collective Action

Community engagement initiatives have the added benefit of promoting social cohesion. Working together in a garden or shopping at a local market creates opportunities for individuals to share knowledge, build relationships, and collaborate on solutions to common challenges. These interactions strengthen community bonds and foster a collective sense of responsibility for creating a healthier, more sustainable food system.

Scaling Impact Through Partnerships

Partnering with local governments, schools, and non-profits can amplify the impact of community engagement efforts. For instance, municipalities can allocate unused land for community gardens, while schools can integrate gardening programs into their curriculums. Collaboration with non-profits can provide funding, expertise, and logistical support, ensuring that these initiatives reach underserved populations.

Hybrid Solutions: Combining the Benefits of Processing with the Nutrition of Whole Foods

Hybrid solutions that blend the advantages of food processing with the nutritional integrity of whole foods represent a promising way forward. These approaches can make healthy options more accessible and convenient without sacrificing quality.

For example, minimally processed foods such as frozen vegetables and vacuum-sealed meals retain much of their nutritional value while extending shelf life and reducing waste. Innovations in fortification and functional foods also enable processed products to address specific dietary needs, such as adding fiber or probiotics to enhance gut health. The goal of hybrid solutions is to strike a balance between convenience and nutrition, meeting the demands of modern life without compromising health (Slavin, 2013).

Policy and Regulation: Advocating for Stricter Controls and Incentives

Policy and regulation play a pivotal role in transforming modern food systems into frameworks that prioritize public health, transparency, and environmental sustainability. By implementing robust regulatory measures and financial incentives, governments and international organizations can influence the production, marketing, and consumption of food, creating a system that better serves people and the planet. This requires a comprehensive approach that addresses additives, labeling, accessibility, and sustainability.

Stricter Controls on Additives and Labeling

One of the most pressing areas for regulatory reform is the use of additives and the clarity of food labeling. The modern food industry relies heavily on chemical additives, preservatives, and flavor enhancers to improve the taste, appearance, and shelf life of products. While some additives are harmless, others have been linked to health risks such as obesity, cardiovascular disease, and hyperactivity in children.

Limiting Harmful Additives

Governments can implement stricter controls to reduce or eliminate harmful additives, including trans fats, high-fructose corn syrup, and artificial dyes. For example, many countries have already banned partially hydrogenated oils, a primary source of industrial trans fats, following their association with heart disease. Expanding these bans to include other controversial additives could significantly improve public health outcomes.

Mandatory Labeling Standards

Transparency in food labeling is equally critical. Regulatory frameworks should require manufacturers to disclose key nutritional information in a clear and accessible manner. For instance, mandatory labeling for added sugars, sodium content, and artificial additives empowers consumers to make informed choices. Countries like Chile and the United Kingdom have introduced front-of-package labeling systems, such as "traffic light" indicators, which use color coding to quickly convey a product's nutritional quality. Studies have shown that these systems can influence purchasing behavior, steering consumers toward healthier options (Campos et al., 2011).

Combating Misleading Claims

Policies must also address deceptive marketing practices. Terms like "natural," "low-fat," or "organic" are often used ambiguously, misleading consumers about a product's healthfulness. Strict definitions and enforcement can prevent these tactics, ensuring that labels accurately reflect the nutritional and environmental attributes of the product.

Incentives for Healthier Options

Economic incentives are powerful tools for shifting consumer behavior and encouraging healthier food production. By making nutritious options more affordable and penalizing unhealthy choices, governments can reshape market demand.

Subsidies for Nutritious Foods

Subsidies for fruits, vegetables, whole grains, and other minimally processed foods can lower prices, making these options more accessible to low-income populations. For example, initiatives like the U.S. Department of Agriculture's (USDA) Fresh Fruit and Vegetable Program provide funding to schools to offer free produce to students, promoting healthier eating habits from an early age.

Taxation on Ultra-Processed Foods

Conversely, imposing taxes on ultra-processed foods high in sugar, unhealthy fats, and sodium can discourage consumption. Countries like Mexico and Hungary have introduced "soda taxes" and levies on junk food, resulting in reduced purchases of sugary beverages and snacks. The revenue generated from these taxes can be reinvested in public health initiatives, such as nutrition education or subsidies for healthier foods.

Supporting Local Food Systems

Incentives can also promote local food production and distribution. Grants or tax breaks for farmers markets, urban agriculture projects, and community-supported agriculture (CSA) programs encourage sustainable practices while improving access to fresh, locally grown produce.

Sustainability Mandates

Aligning food systems with environmental goals is essential for addressing climate change, resource scarcity, and biodiversity loss. Policies promoting sustainability can drive innovation and accountability across the food industry.

Reducing Food Waste

Globally, nearly one-third of all food produced is wasted, contributing to significant greenhouse gas emissions and resource depletion. Governments can implement policies to reduce waste at every stage of the supply chain, from production to consumer behavior. For example, France has mandated that supermarkets donate unsold food to charities, while South Korea has introduced strict food waste recycling laws.

Eco-Friendly Packaging

Incentivizing the use of biodegradable, compostable, or reusable packaging can significantly reduce plastic pollution. Policies requiring manufacturers to adopt sustainable packaging materials or imposing fees on single-use plastics encourage innovation and reduce environmental harm.

Investing in Sustainable Technologies

Public funding for research and development in sustainable food production, such as regenerative agriculture and precision fermentation, can accelerate progress. Governments can also offer grants or low-interest loans to companies developing technologies that minimize environmental impact, such as lab-grown meat or vertical farming.

Incentivizing Climate-Resilient Practices

Climate change poses significant risks to food security, making it essential to promote climate-resilient agricultural practices. Policies supporting crop diversification, soil conservation, and water-efficient irrigation systems can help mitigate these risks while preserving natural ecosystems.

Reclaiming control over modern food systems demands bold, collective action that balances innovation with integrity, accessibility with accountability, and progress with preservation. By redefining progress to prioritize health, sustainability, and equity, we can transform food systems into drivers of well-being rather than contributors to crises. Harnessing technological breakthroughs, empowering informed consumer choices, and enforcing robust policies are not just strategies—they are necessities. Together, these efforts can forge a future where convenience and nutrition coexist, cultural heritage thrives, and environmental stewardship becomes a cornerstone of how we nourish the world. It is a vision of progress that serves not just today's needs but secures a healthier, more sustainable legacy for generations to come.

Chapter 17

Living with Balance
The Pragmatic Approach

As the modern world grapples with the challenges of health crises, environmental degradation, and cultural shifts in food systems, finding a balanced approach to eating is more important than ever. Living with balance doesn't require extreme measures or abandoning the convenience of processed foods altogether. Instead, it calls for pragmatic strategies that allow individuals, families, and communities to make better choices within the constraints of modern life. This chapter explores how to achieve balance in daily diets, the role of communities in supporting healthier food systems, and the potential for a future that aligns health, sustainability, and convenience.

The Realistic Diet: Finding a Balance Between Processed and Unprocessed Foods

A realistic diet acknowledges that while avoiding processed foods entirely may not be practical for many, it is possible to minimize their role in daily life by focusing on whole and minimally processed alternatives. Processed foods are often integrated into modern lifestyles due to their convenience, affordability, and accessibility.

Striking a balance involves recognizing the spectrum of processing and making informed decisions about which foods to prioritize and which to limit.

Understanding the Spectrum of Processing

Processed foods exist on a spectrum, from minimally processed items that retain most of their natural nutritional value to ultra-processed products that are significantly altered and often laden with unhealthy additives. At one end are ultra-processed foods like sugary beverages, packaged snacks, and ready-to-eat meals. These are typically high in added sugars, unhealthy fats, and sodium while offering little nutritional benefit. Regular consumption of such foods has been linked to a higher risk of chronic diseases like obesity, diabetes, and cardiovascular issues (Monteiro et al., 2019).

At the other end of the spectrum are minimally processed foods, such as frozen vegetables, canned beans, and whole-grain bread. These foods undergo minimal processing to enhance shelf life or usability while preserving their nutrient content. For example, frozen vegetables are picked at peak ripeness and quickly frozen to retain vitamins and minerals. Including these in a diet provides convenience without compromising nutrition.

A balanced diet incorporates minimally processed foods as practical and nutritious choices, using ultra-processed items as occasional treats rather than staples. For example, preparing a frozen vegetable stir-fry with whole-grain rice is both convenient and nourishing, offering a quick meal that aligns with health goals. Understanding this spectrum empowers individuals to make choices that suit their lifestyle and health needs.

Incorporating Flexibility

Rigid dietary rules can often lead to frustration, burnout, or feelings of guilt when they are difficult to maintain. A balanced diet emphasizes flexibility and the overall pattern of eating, allowing individuals to focus on long-term habits rather than striving for perfection at every meal.

One practical framework is the 80/20 rule, where 80% of meals prioritize whole, nutrient-dense foods, and 20% allow for indulgences or convenience items. This approach reduces the pressure to adhere to an unrealistic ideal while encouraging mindfulness about food choices. For instance, enjoying a slice of pizza during a family gathering or indulging in a dessert occasionally is perfectly acceptable when balanced by predominantly nutritious meals.

Flexibility also acknowledges the realities of varying schedules, budgets, and preferences. By adapting to these factors without compromising overall goals, individuals are more likely to sustain healthy eating habits over time. For example, a busy week might include more reliance on quick options like whole-grain sandwiches or pre-washed salad greens, while a calmer weekend might allow for experimenting with home-cooked meals.

Practical Steps for Families: Meal Planning, Cooking Techniques, and Making Better Choices on a Budget

Families face unique challenges when trying to maintain a balanced diet, including limited time, picky eaters, and tight budgets. Practical strategies like meal planning, simple cooking techniques, and budget-friendly shopping can help households improve their diets without feeling overwhelmed.

Meal Planning

Meal planning is a foundational strategy for balanced eating that helps families reduce stress, minimize food waste, and ensure variety. By creating a weekly menu, families can plan meals that incorporate fresh, seasonal ingredients while avoiding the last-minute temptation to opt for fast food or heavily processed options.

Designating specific themes for different days, such as "Meatless Monday," "Taco Tuesday," or "Soup Saturday," simplifies decision-making and provides structure. For example, "Meatless Monday" could include dishes like lentil curry or vegetable stir-fry, encouraging the family to experiment with plant-based meals while reaping their nutritional and environmental benefits.

Batch cooking is another effective method, allowing families to prepare larger quantities of meals that can be stored and reheated during busy nights. Dishes like soups, casseroles, or chili are particularly suited for this approach, as they maintain their flavor and nutritional value when stored. Having pre-made meals on hand reduces reliance on fast food and ultra-processed options, making it easier to stick to a balanced diet.

Shopping Smartly on a Budget

Meal planning also helps families make better choices on a budget by reducing impulse purchases and maximizing the value of each grocery trip. Staples like rice, beans, oats, and frozen vegetables are cost-effective and versatile, forming the base of many nutritious meals. Buying in bulk, choosing store brands, and taking advantage of discounts or sales can further stretch the grocery budget.

For example, a family can plan a week's worth of meals around budget-friendly items like roasted sweet potatoes, brown rice, and canned chickpeas, supplemented by seasonal produce. Incorporating cost-effective sources of protein, such as eggs, lentils, or tofu, provides variety without breaking the bank.

Meal planning also reduces food waste by encouraging the use of all purchased ingredients. For instance, leftover roasted vegetables from one dinner can be repurposed into a frittata for breakfast or a salad for lunch, ensuring nothing goes to waste.

Cooking Techniques

Teaching basic cooking skills is essential for reducing dependence on processed foods. Simple techniques such as roasting vegetables, preparing lean proteins, and making homemade sauces can transform basic ingredients into flavorful meals. For instance, swapping store-bought pasta sauce for a homemade version made with canned tomatoes, garlic, and herbs not only reduces added sugars but also saves money.

Cooking as a family can also be a bonding experience, teaching children valuable skills and encouraging them to try new foods. Engaging kids in tasks like washing vegetables or stirring soups helps them feel invested in the meal.

Making Better Choices on a Budget

Healthy eating doesn't have to break the bank. Families can stretch their budgets by prioritizing cost-effective staples like rice, beans, oats, and frozen vegetables. Buying in bulk, choosing store brands, and taking advantage of sales on fresh produce can further reduce costs.

For those with limited access to fresh foods, canned and frozen options are excellent alternatives. Look for products with no added sugars or sodium, such as unsweetened canned fruits or low-sodium beans. Planning meals around affordable, nutrient-dense ingredients ensures both health and cost-effectiveness.

The Role of Communities: How Local Food Systems and Farmers' Markets Can Play a Part

Communities play a vital role in fostering balanced eating by developing and supporting local food systems, creating a stronger connection between consumers and the food they eat, and making fresh, nutritious options more accessible. Local food systems provide an opportunity for sustainable practices, community engagement, and improved food security. These systems offer an alternative to large-scale, industrial food production, creating opportunities for healthier eating and environmental stewardship.

Farmers' Markets and Local Produce

Farmers' markets serve as a direct link between consumers and local farmers, providing fresh, seasonal produce that might otherwise be unavailable in supermarkets. These markets help strengthen the local economy by allowing small-scale farmers to sell their products directly to consumers, eliminating the middlemen involved in industrial supply chains. This system also reduces the carbon footprint associated with transporting food over long distances, as consumers are buying locally grown produce.

Additionally, farmers' markets often feature a range of organic and sustainable produce, which provides consumers with healthier options free from harmful pesticides or artificial fertilizers. These markets often go beyond the simple exchange of goods; they also serve as educational platforms, offering workshops on topics like nutrition, sustainable farming, and seasonal eating. Consumers can learn about how their food is grown, which in turn fosters more mindful purchasing decisions and encourages support for local agriculture.

Many farmers' markets now participate in programs like the Supplemental Nutrition Assistance Program (SNAP), and some even offer matching programs or discounts to low-income families, making fresh produce more accessible to those who might otherwise be unable to afford it. This initiative helps reduce food insecurity while promoting healthier diets in communities that may lack easy access to affordable, nutritious food (Gundersen & Ziliak, 2015). As such, farmers' markets not only serve as a hub for healthier food options but also act as important community spaces where individuals can build connections with local food producers and other like-minded individuals.

Community Gardens

Community gardens provide a transformative way for residents to directly engage with food production, offering opportunities to grow their own fruits, vegetables, and herbs. These gardens promote self-reliance and reduce dependence on store-bought, processed foods by encouraging people to cultivate fresh produce in their own neighborhoods. Community gardens are particularly important in urban areas, where access to affordable, fresh food can be limited, and they offer an opportunity to reclaim vacant lots or underutilized urban spaces for productive use.

These gardens often have educational components, teaching individuals about sustainable agricultural practices such as composting, water conservation, and organic farming. By participating in community gardens, individuals gain a deeper understanding of where their food comes from, fostering a greater appreciation for fresh produce and the work involved in growing it. They also contribute to biodiversity by providing habitats for pollinators, which are crucial for healthy ecosystems.

Additionally, community gardens can improve social cohesion, as they bring together people from diverse backgrounds to work collaboratively towards a common goal. This shared effort promotes a sense of ownership and community pride while creating spaces for learning and connection. The health benefits are also substantial, as individuals who participate in gardening activities often enjoy greater physical activity and better access to fresh, nutrient-dense foods.

Food Co-ops and CSAs

Food cooperatives (co-ops) and Community Supported Agriculture (CSA) programs are excellent ways for communities to access fresh, locally sourced ingredients at reasonable prices while supporting sustainable farming practices. Food co-ops are member-owned grocery stores that prioritize locally grown, organic, and ethically sourced food, often with a focus on reducing food waste and promoting environmental sustainability. Co-ops allow members to purchase food at discounted prices and offer a sense of community, as they are often run by volunteers or with strong community involvement.

Community Supported Agriculture (CSA) programs allow individuals or families to subscribe to a local farm, receiving a regular share of seasonal produce. This not only guarantees access to fresh, high-quality food but also encourages families to consume more fruits and vegetables, many of which they may not otherwise purchase in a traditional grocery store. CSAs promote seasonal eating, which supports a more sustainable food system by reducing the need for long-distance food transport and encouraging consumption of what is locally available.

Both food co-ops and CSAs create stronger connections between food producers and consumers. They foster a deeper understanding of where food comes from and the processes involved in its production, allowing consumers to make more informed choices. These systems also support local farmers by ensuring they have a steady, committed customer base, helping to maintain small-scale farming operations and making them more economically viable.

Looking to the Future: Can We Create a Food System That Supports Health, Sustainability, and Modern Convenience?

The future of food lies in creating a system that supports the needs of both the planet and its inhabitants while adapting to the demands of modern lifestyles. Balancing health, sustainability, and convenience is no small task, but it is increasingly clear that a multifaceted approach is required to meet these goals.

Advancements in technology, such as lab-grown meat and vertical farming, offer promising solutions to food production's environmental challenges. These innovations could help create sustainable alternatives to traditional animal farming, reducing land use, water consumption, and greenhouse gas emissions while providing the convenience and accessibility that modern consumers demand. As urban populations grow, solutions like vertical farming in urban centers could also offer localized food production, reducing transportation costs and food waste.

At the same time, public policy plays a crucial role in shaping this future. Governments must incentivize practices that align food production with sustainability goals, encourage transparency in food labeling, and promote access to healthy, affordable options for all. Strengthening local food systems and expanding initiatives such as food co-ops, CSA programs, and farmers' markets will be essential for creating communities that are not only more food-secure but also more connected to the source of their food.

Creating a food system that promotes health, sustainability, and modern convenience requires collaboration between consumers, communities, and policymakers. By supporting local food systems, encouraging innovative solutions, and ensuring that sustainable options are accessible to everyone, we can build a food future that nourishes people and the planet alike.

Technological Advancements

Emerging technologies, such as lab-grown meats, vertical farming, and AI-driven agricultural systems, hold promise for creating a food system that aligns with environmental goals. These innovations can reduce resource use, enhance nutritional quality, and provide scalable solutions for feeding growing populations (Post et al., 2020).

Policy Reform

Governments must play an active role in reshaping food systems through subsidies for sustainable farming, regulations on harmful additives, and incentives for reducing food waste. Policies that prioritize access to healthy foods for all income levels can help bridge the gap between affordability and nutrition.

Cultural Shifts

Creating a healthier food system also requires cultural shifts that value quality over convenience. Public education campaigns, community engagement, and social media movements can raise awareness about the benefits of balanced eating, encouraging individuals to prioritize health and sustainability in their choices.

Finally, building a food system that balances health, sustainability, and convenience is not only possible but essential for the well-being of individuals, communities, and the planet. By strengthening local food systems, embracing innovative technologies, and fostering a culture of informed consumer choices, we can create a future where nutritious, sustainable food is accessible to all. This collective effort—driven by innovation, education, and policy—holds the potential to redefine the way we nourish ourselves and our world, ensuring a healthier, more sustainable future for generations to come.

Chapter 18

Toward a Healthier Future

The journey toward a healthier future requires collaboration, innovation, and an unwavering commitment to creating a balanced food system that nourishes both people and the planet. Throughout this book, we have explored the complexities of modern food systems, the challenges posed by processed foods, and the urgent need for systemic change. From the individual choices we make every day to the influence of the food industry and the policies that shape the food landscape, each element plays a crucial role in transforming our diets and the food system as a whole.

This chapter serves as the culmination of the conversations we have had about food systems, health, sustainability, and convenience. It is a call to action for individuals, communities, industries, and policymakers to work together in creating a future where processed foods are balanced with whole, nutritious options that support health, cultural traditions, and environmental sustainability.

The Dual Responsibility of Individuals and the Food Industry

As we have seen, the path to healthier eating is shared by individuals and the food industry. Consumers play a crucial role in making informed choices, becoming more food literate, and understanding the nutritional value of the foods they eat. Food literacy encompasses everything from reading food labels to understanding the environmental impact of dietary choices, allowing individuals to make decisions that benefit both their health and the planet.

However, it is important to acknowledge that individual responsibility is often shaped by the environment and the food options available. The food industry, as a central player in the global food system, bears a significant responsibility to offer healthier products and reformulate existing ones to improve their nutritional value. By reducing added sugars, unhealthy fats, and artificial ingredients, and focusing on transparency in labeling, food companies can make healthier choices more accessible to the masses. Reformulating processed foods to offer healthier alternatives that align with public health goals is not only possible but necessary for long-term change. The food industry must prioritize these changes, recognizing that the health of consumers and the health of the planet must go hand in hand.

Advocacy for Policy Changes, Better Labeling, and Food Education

Advocacy for policy changes, better labeling, and widespread food education is essential to create an environment where healthier eating is the norm. Governments must play an active role in shaping the future of food by implementing policies that encourage healthier food production and consumption. Subsidizing healthy foods like fruits, vegetables, and whole grains can make nutritious options more affordable for all, while taxing ultra-processed foods high in sugar and unhealthy fats can reduce consumption and drive demand for healthier alternatives.

Food labeling plays a critical role in empowering consumers to make informed decisions. Clear, honest labeling that discloses nutritional content, portion sizes, and the presence of harmful additives is essential for building consumer trust. Systems like front-of-package traffic light labels, which quickly convey a product's nutritional value, can help individuals make better choices at a glance. In parallel, mandatory education initiatives about food and nutrition, both in schools and through public health campaigns, can raise awareness about the long-term health benefits of balanced eating, making healthy choices an accessible and achievable goal for everyone.

A Vision for the Future of Food

The ultimate vision for the future is a world where processed foods are not demonized but are thoughtfully integrated into a diet that prioritizes balance, health, and transparency. In this future, convenience foods are balanced with whole, nutritious options, making it easier for individuals and families to nourish themselves without sacrificing taste, convenience, or enjoyment. Transparency and accountability within the food industry play a pivotal role, ensuring that consumers have access to accurate information about ingredients, nutritional value, and environmental impact. By addressing deceptive marketing and hidden health risks, the food industry can rebuild trust and empower individuals to make informed choices.

Innovation and technology hold immense potential to reshape the food landscape. Advancements in food processing, lab-grown meats, and sustainable packaging lead the charge in creating healthier, more eco-friendly options. These innovations are complemented by a resurgence in local food systems and community-driven initiatives, such as farmers' markets, community gardens, and food cooperatives. Together, they empower individuals to reclaim control over their food choices, providing access to fresh, healthy options that support both individual health and the local economy. Education and awareness are essential in this journey, equipping people with the knowledge to navigate the complexities of the modern food system and advocate for meaningful change.

At the heart of this vision lies a redefinition of what it means to "eat well" in the modern world. A balanced food system should allow for the enjoyment of foods that are not only convenient but also nourishing, sustainable, and ethically produced. This system must also celebrate the rich cultural diversity of food, preserving traditional practices while embracing modern solutions. Food is deeply intertwined with identity and community, and a sustainable future must respect and honor these connections.

Achieving this vision requires cooperation from all sectors of society—from individual choices to industry practices to policy implementation. Governments play a critical role in enacting policies that make healthy foods more accessible, support local food systems, and hold corporations accountable for ethical and sustainable practices. At the same time, individuals can drive change by demanding greater transparency and accountability from the food industry, pushing for policies that prioritize public health over profits.

Communities also have the power to lead this transformation by fostering local food initiatives, promoting education about nutrition and sustainability, and providing resources to support equitable access to healthy foods. By empowering individuals to take charge of their food systems, communities can create a ripple effect that benefits health, the environment, and the economy.

The road ahead will not be easy, but it is possible. By balancing the benefits of convenience with the need for nutrition, sustainability, and equity, we can create a food system that promotes health for all, respects the environment, and celebrates cultural diversity. This transformation requires commitment, collaboration, and hope.

Through collective efforts, we can build a food system that nourishes both people and the planet for generations to come. By reclaiming control over the way, we eat and embracing the challenges and opportunities before us, we can create a future where food is a source of joy, health, and connection—not just for ourselves but for future generations.

Bibliography

Barboza, L. G., Vethaak, A. D., Lavorante, B. R., Lundebye, A. K., & Guilhermino, L. (2018). Marine microplastic debris: An emerging issue for food security, food safety, and human health. Marine Pollution Bulletin, 133, 336–348. https://doi.org/10.1016/j.marpolbul.2018.05.047

Bray, G. A., Nielsen, S. J., & Popkin, B. M. (2004). Consumption of high-fructose corn syrup in beverages may play a role in the epidemic of obesity. American Journal of Clinical Nutrition, 79(4), 537–543. https://doi.org/10.1093/ajcn/79.4.537

Campos, S., Doxey, J., & Hammond, D. (2011). Nutrition labels on pre-packaged foods: A systematic review. Public Health Nutrition, 14(8), 1496–1506. https://doi.org/10.1017/S1368980010003290

Cani, P. D., Bibiloni, R., Knauf, C., Waget, A., Neyrinck, A. M., Delzenne, N. M., & Burcelin, R. (2008). Changes in gut microbiota control metabolic endotoxemia-induced inflammation in high-fat diet-induced obesity and diabetes in mice. Diabetes, 57(6), 1470–1481. https://doi.org/10.2337/db07-1403

Chandon, P., & Wansink, B. (2012). The biasing health halos of fast-food restaurant health claims: Lower calorie estimates and higher side-dish consumption intentions. Journal of Consumer Research, 34(3), 301–314. https://doi.org/10.1086/519499

Chassaing, B., Koren, O., Goodrich, J. K., Poole, A. C., Srinivasan, S., Ley, R. E., & Gewirtz, A. T. (2015). Dietary emulsifiers impact the mouse gut microbiota promoting colitis and metabolic syndrome. Nature, 519(7541), 92–96. https://doi.org/10.1038/nature14232

Chopra, M., Galbraith, S., & Darnton-Hill, I. (2002). A global response to a global problem: The epidemic of overnutrition.

Bulletin of the World Health Organization, 80(12), 952–958. https://www.who.int

Clark, M. A., Springmann, M., Hill, J., & Tilman, D. (2019). Multiple health and environmental impacts of foods. Proceedings of the National Academy of Sciences, 116(46), 23357-23362. https://doi.org/10.1073/pnas.1906908116

Colchero, M. A., Salgado, J. C., Unar-Munguía, M., Hernández-Ávila, M., & Monsiváis, P. (2016). Beverage purchases from stores in Mexico under the excise tax on sugar-sweetened beverages: Observational study. BMJ, 352, h6704. https://doi.org/10.1136/bmj.h6704

Deehan, E. C., & Walter, J. (2017). The fiber gap and the disappearing gut microbiome: Implications for human nutrition. Trends in Endocrinology & Metabolism, 28(5), 399-412. https://doi.org/10.1016/j.tem.2017.02.003

Drewnowski, A., & Almiron-Roig, E. (2010). Human perceptions and preferences for fat-rich foods. Current Opinion in Clinical Nutrition & Metabolic Care, 13(6), 625–632. https://doi.org/10.1097/MCO.0b013e32834019a4

Drewnowski, A., & Darmon, N. (2005). The economics of obesity: Dietary energy density and energy cost. American Journal of Clinical Nutrition, 82(1), 265S–273S. https://doi.org/10.1093/ajcn/82.1.265S

Fardet, A. (2016). Minimally processed foods are more satiating and less hyperglycemic than ultra-processed foods: A preliminary study with 98 ready-to-eat foods. Food & Function, 7(5), 2338–2346. https://doi.org/10.1039/c6fo00107f

Fasano, A. (2012). Leaky gut and autoimmune diseases. Clinical Reviews in Allergy & Immunology, 42(1), 71–78. https://doi.org/10.1007/s12016-011-8291-x

Foley, J. A., Ramankutty, N., Brauman, K. A., Cassidy, E. S., Gerber, J. S., Johnston, M., ... & Zaks, D. P. M. (2011). Solutions for a cultivated planet. Nature, 478(7369), 337-342. https://doi.org/10.1038/nature10452

Food and Agriculture Organization of the United Nations (FAO). (2021). The state of food and agriculture: Making agri-food systems more resilient. Rome: FAO.

Friedman, M. (2003). Chemistry, biochemistry, and safety of acrylamide. Journal of Agricultural and Food Chemistry, 51(16), 4504–4526. https://doi.org/10.1021/jf030204+

Gall, S. C., & Thompson, R. C. (2015). The impact of debris on marine life. Marine Pollution Bulletin, 92(1-2), 170-179. https://doi.org/10.1016/j.marpolbul.2014.12.041

Garnett, T. (2011). Where are the best opportunities for reducing greenhouse gas emissions in the food system (including the food chain)? Food Policy, 36(S1), S23-S32. https://doi.org/10.1016/j.foodpol.2010.10.010

Gundersen, C., & Ziliak, J. P. (2015). Food insecurity and health outcomes. Health Affairs, 34(11), 1830–1839. https://doi.org/10.1377/hlthaff.2015.0645

Harris, J. L., Schwartz, M. B., & Brownell, K. D. (2019). Marketing foods to children and adolescents: Licensed characters and other promotions on packaged foods in the supermarket. Public Health Nutrition, 12(3), 414-421. https://doi.org/10.1017/S1368980008002262

Hawkes, C. (2006). Uneven dietary development: Linking the policies and processes of globalization with the nutrition transition, obesity, and diet-related chronic diseases. Globalization and Health, 2(4). https://doi.org/10.1186/1744-8603-2-4

He, F. J., & MacGregor, G. A. (2010). Reducing population salt intake worldwide: From evidence to implementation. Progress in Cardiovascular Diseases, 52(5), 363–382. https://doi.org/10.1016/j.pcad.2009.12.006

He, F. J., Pombo-Rodrigues, S., & MacGregor, G. A. (2014). Salt reduction in England from 2003 to 2011: Its relationship to blood pressure, stroke, and ischemic heart disease mortality. BMJ Open, 4(4), e004549. https://doi.org/10.1136/bmjopen-2013-004549

He, F. J., & MacGregor, G. A. (2007). Salt, blood pressure and cardiovascular disease. Current Opinion in Cardiology, 22(4), 298–305. https://doi.org/10.1097/MCO.0b013e32814a55f0

Heaney, R. P. (2001). Factors influencing the measurement of bioavailability, taking calcium as a model. Journal of Nutrition, 131(4), 1344S-1348S. https://doi.org/10.1093/jn/131.4.1344S

Hu, F. B., Satija, A., & Rimm, E. B. (2014). Diet and cardiovascular disease: The role of whole and refined grains. Journal of the American College of Cardiology, 64(10), 962-970. https://doi.org/10.1016/j.jacc.2014.06.047

IBISWorld. (2023). Food & beverage manufacturing in the US - Market research report. Retrieved from https://www.ibisworld.com

Institute of Food Technologists (IFT). (2018). Food processing and its impact on the environment and health. IFT Press.

Jambeck, J. R., Geyer, R., Wilcox, C., Siegler, T. R., Perryman, M., Andrady, A., & Law, K. L. (2015). Plastic waste inputs from land into the ocean. Science, 347(6223), 768-771. https://doi.org/10.1126/science.1260352

Kaur, A., Scarborough, P., & Rayner, M. (2017). A systematic review, and meta-analyses, of the impact of health-related claims on dietary choices. International Journal of Behavioral Nutrition and Physical Activity, 14(1), 93. https://doi.org/10.1186/s12966-017-0548-1

Knorr, D., Ade-Omowaye, B. I., & Heinz, V. (2011). Nutritional improvement of plant foods by non-thermal processing. Proceedings of the Nutrition Society, 61(2), 311-318. https://doi.org/10.1079/PNS2002152

Kris-Etherton, P. M., Hecker, K. D., Bonanome, A., Coval, S. M., Binkoski, A. E., Hilpert, K. F., ... & Etherton, T. D. (2013). Bioactive compounds in foods: Their role in the prevention of cardiovascular disease and cancer. American Journal of Medicine, 113(9), 71S-88S. https://doi.org/10.1016/j.amjmed.2013.05.001

Kumar, R., & Sharma, A. (2020). Industrial effluents and their impact on water quality and ecosystems. Environmental Chemistry Letters, 18(2), 543-560. https://doi.org/10.1007/s10311-020-01016-4

Lahou, E., Uyttendaele, M., & De Loy-Hendrickx, A. (2017). Impact of food processing on the safety and quality of foods. Food Control, 78, 143–151. https://doi.org/10.1016/j.foodcont.2017.02.024

Lassale, C., Gunter, M. J., Romaguera, D., Peelen, L. M., van der Schouw, Y. T., Beulens, J. W., & Riboli, E. (2018). Diet quality and risk of obesity and metabolic syndrome: A systematic review and meta-analysis of prospective studies. PLoS Medicine, 15(6), e1002715. https://doi.org/10.1371/journal.pmed.1002715

Liu, R. H. (2013). Health-promoting components of fruits and vegetables in the diet. Advances in Nutrition, 4(3), 384S-392S. https://doi.org/10.3945/an.112.003517

Lyon, T. P., & Maxwell, J. W. (2011). Greenwash: Corporate environmental disclosure under threat of audit. Journal of Economics & Management Strategy, 20(1), 3–41. https://doi.org/10.1111/j.1530-9134.2010.00282.x

Magnuson, B. A., Burdock, G. A., Doull, J., Kroes, R. M., Marsh, G. M., Pariza, M. W., Spencer, P. S., Waddell, W. J., Walker, R., & Williams, G. M. (2007). Aspartame: A safety evaluation based on current use levels, regulations, and toxicological and epidemiological studies. Critical Reviews in Toxicology, 37(8), 629–727. https://doi.org/10.1080/10408440701516184

Meijaard, E., Garcia-Ulloa, J., Sheil, D., Wich, S. A., Carlson, K. M., Juffe-Bignoli, D., & Brooks, T. M. (2018). Oil palm and biodiversity: A situation analysis by the IUCN Oil Palm Task Force. International Union for Conservation of Nature. https://doi.org/10.2305/IUCN.CH.2018.11.en

Micha, R., Peñalvo, J. L., Cudhea, F., Imamura, F., Rehm, C. D., & Mozaffarian, D. (2017). Association between dietary factors and mortality from heart disease, stroke, and type 2 diabetes in the United States. JAMA, 317(9), 912-924. https://doi.org/10.1001/jama.2017.0947

Monteiro, C. A., Cannon, G., Levy, R. B., Moubarac, J.-C., Jaime, P., Martins, A. P., Canella, D., Louzada, M., & Parra, D. (2018). Ultra-processed foods: What they are and how to identify them. Public Health Nutrition, 21(1), 7–17. https://doi.org/10.1017/S1368980018003762

Moss, M. (2013). Salt sugar fat: How the food giants hooked us. Random House.

Mozaffarian, D., Angell, S. Y., Lang, T., & Rivera, J. A. (2018). Role of government policy in nutrition—Barriers to and opportunities for healthier eating. BMJ, 361, k2426. https://doi.org/10.1136/bmj.k2426

Muller, L., Lacroix, A., & Ruffieux, B. (2017). Environmental claims in food advertising: Investigating the impact of nature-evoking elements. Appetite, 108, 464–472. https://doi.org/10.1016/j.appet.2016.11.010

Nestle, M. (2013). Food politics: How the food industry influences nutrition and health. University of California Press.

O'Donnell, M. J., Mente, A., Rangarajan, S., McQueen, M. J., Wang, X., Liu, L., ... & Yusuf, S. (2014). Urinary sodium and potassium excretion, mortality, and cardiovascular events. New England Journal of Medicine, 371(7), 612–623. https://doi.org/10.1056/NEJMoa1311889

Okrent, A. M., & Alston, J. M. (2011). The effects of farm commodity and retail food policies on obesity and economic welfare in the United States. American Journal of Agricultural Economics, 93(2), 378-384. https://doi.org/10.1093/ajae/aaq104

Pollan, M. (2008). In defense of food: An eater's manifesto. Penguin Press.

Popkin, B. M. (2006). Global nutrition dynamics: The world is shifting rapidly toward a diet linked with noncommunicable diseases. The American Journal of Clinical Nutrition, 84(2), 289–298. https://doi.org/10.1093/ajcn/84.2.289

Popkin, B. M., Adair, L. S., & Ng, S. W. (2020). Global nutrition transition and the pandemic of obesity in developing countries. Nutrition Reviews, 68(1), 3-21. https://doi.org/10.1111/j.1753-4887.2009.00290.x

Post, M. J., Levenberg, S., Kaplan, D. L., Genovese, N., Fu, J., Bryant, C. J., ... & Fleischmann, A. M. (2020). Scientific, sustainability, and regulatory challenges of cultured meat. Nature Food, 1(7), 403–415. https://doi.org/10.1038/s43016-020-0112-z

Powell, L. M., Harris, J. L., & Fox, T. (2013). Food marketing expenditures aimed at youth: Putting the numbers in context. American Journal of Preventive Medicine, 45(4), 453-460. https://doi.org/10.1016/j.amepre.2013.06.003

Ratti, C. (2001). Hot air and freeze-drying of high-value foods: A review. Journal of Food Engineering, 49(4), 311–319. https://doi.org/10.1016/S0260-8774(00)00228-4

Roberto, C. A., & Khandpur, N. (2014). Improving the design of nutrition labels to promote healthier food choices and reasonable portion sizes. International Journal of Obesity, 38(S1), S25–S33. https://doi.org/10.1038/ijo.2014.86

Rubin, B. S. (2011). Bisphenol A: An endocrine disruptor with widespread exposure and multiple effects. The Journal of Steroid Biochemistry and Molecular Biology, 127(1-2), 27–34. https://doi.org/10.1016/j.jsbmb.2011.05.002

Rujnić-Sokele, M., & Pilipović, A. (2017). Challenges and opportunities of biodegradable plastics: A mini-review. Waste Management & Research, 35(2), 132-140. https://doi.org/10.1177/0734242X16683272

Sacks, G., Rayner, M., & Swinburn, B. (2015). Impact of front-of-pack 'traffic-light' nutrition labels on consumer food purchases in the UK. Health Promotion International, 30(2), 224-234. https://doi.org/10.1093/heapro/dau110

Simopoulos, A. P. (2002). Omega-3 fatty acids in inflammation and autoimmune diseases. Journal of the American College of Nutrition, 21(6), 495–505. https://doi.org/10.1080/07315724.2002.10719248

Singh, S., Gamlath, S., & Wakeling, L. (2007). Nutritional aspects of food extrusion: A review. International Journal of Food Science & Technology, 42(8), 916–929. https://doi.org/10.1111/j.1365-2621.2006.01309.x

Slavin, J. L. (2013). Fiber and prebiotics: Mechanisms and health benefits. Nutrients, 5(4), 1417-1435. https://doi.org/10.3390/nu5041417

Stern, P. C., & Dietz, T. (2002). Consumer perceptions of environmental claims on product labels. Environment and Behavior, 34(2), 150–171. https://doi.org/10.1177/0013916502034002002

Stevens, L. J., Burgess, J. R., Stochelski, M. A., & Kuczek, T. (2013). Amounts of artificial food colors in commonly consumed beverages and potential behavioral implications for consumption in children. Clinical Pediatrics, 52(2), 133–140. https://doi.org/10.1177/0009922812474536

Strazzullo, P., D'Elia, L., Kandala, N.-B., & Cappuccio, F. P. (2009). Salt intake, stroke, and cardiovascular disease: Meta-analysis of prospective studies. BMJ, 339, b4567. https://doi.org/10.1136/bmj.b4567

Suez, J., Korem, T., Zeevi, D., Zilberman-Schapira, G., Thaiss, C. A., Maza, O., Israeli, D., & Elinav, E. (2014). Artificial sweeteners induce glucose intolerance by altering the gut microbiota. Nature, 514(7521), 181–186. https://doi.org/10.1038/nature13793

Tilman, D., Cassman, K. G., Matson, P. A., Naylor, R., & Polasky, S. (2001). Agricultural sustainability and intensive production practices. Nature, 418(6898), 671-677. https://doi.org/10.1038/nature01014

Teufel, B., & Brown, K. (2020). The health halo effect: Influence of food labeling and marketing on consumer perception. Food Research International, 132, 109035. https://doi.org/10.1016/j.foodres.2020.109035

Vandenberg, L. N., Maffini, M. V., Sonnenschein, C., Rubin, B. S., & Soto, A. M. (2012). Bisphenol-A and the great divide: A review of controversies in the field of endocrine disruption. Endocrine Reviews, 30(1), 75–95. https://doi.org/10.1210/er.2008-0021

Walker, R. E., Keane, C. R., & Burke, J. G. (2010). Disparities and access to healthy food in the United States: A review of food deserts literature. Health & Place, 16(5), 876-884. https://doi.org/10.1016/j.healthplace.2010.04.013

Willett, W., Rockström, J., Loken, B., Springmann, M., Lang, T., Vermeulen, S., ... & Murray, C. J. L. (2019). Food in the Anthropocene: the EAT–Lancet Commission on healthy diets from sustainable food systems. The Lancet, 393(10170), 447-492. https://doi.org/10.1016/S0140-6736(18)31788-4

Willett, W. C., & Stampfer, M. J. (2013). Rebuilding the food pyramid. Scientific American, 288(1), 64–71. https://doi.org/10.1038/scientificamerican0401-64

Wilcove, D. S., & Koh, L. P. (2010). Addressing the threats to biodiversity from oil-palm agriculture. Biodiversity and Conservation, 19(4), 999-1007. https://doi.org/10.1007/s10531-010-9789-8

Wolfson, J. A., & Bleich, S. N. (2015). Is cooking at home associated with better diet quality or weight-loss intention? Public Health Nutrition, 18(8), 1397–1406. https://doi.org/10.1017/S1368980014001943

Supplemental Materials

Glossary: Common Chemical Terms and Food Additives

Understanding the language of food science is essential to navigating modern diets. Below is a glossary of commonly encountered chemical terms and food additives that you might find on food labels. This glossary helps demystify some of the ingredients used in processed foods and provides clarity on their function and potential effects on health.

- **Additive**: Substances added to food during processing to enhance its flavor, texture, appearance, or shelf-life. Examples include preservatives, colorings, and flavorings.

- **Artificial Sweeteners**: Chemical compounds that mimic the taste of sugar but are usually lower in calories. Common examples include aspartame, saccharin, and sucralose. While these additives provide sweetness without calories, concerns over their long-term health effects persist.

- **BHA (Butylated Hydroxyanisole)**: A synthetic antioxidant used to preserve fats and oils in processed foods. It has been associated with potential health risks, including cancer in animal studies, although the evidence in humans is inconclusive.

- **High-Fructose Corn Syrup (HFCS)**: A sweetener made from corn starch that is commonly used in soft drinks, candies, and processed foods. It has been linked to obesity and other metabolic issues when consumed in excess.

- **Monosodium Glutamate (MSG)**: A flavor enhancer used in many processed foods, particularly in savory products like soups, snacks, and sauces. Although considered safe by regulatory authorities, some individuals may experience sensitivity to it, known as "Chinese restaurant syndrome."

- **Partially Hydrogenated Oils**: Oils that have undergone partial hydrogenation to increase shelf life and improve texture. They contain trans fats, which are linked to cardiovascular disease, and are now being phased out in many countries due to their health risks.

- **Sodium Benzoate**: A common preservative used in acidic foods and beverages like sodas, fruit juices, and pickles. While generally recognized as safe, sodium benzoate can cause allergic reactions in some individuals and may form benzene, a carcinogen, when exposed to heat and light.

- **Citric Acid**: A natural preservative found in citrus fruits, used to add tartness to foods and extend shelf life. It is commonly found in juices, candies, and soft drinks and is considered safe for consumption.

- **Caramel Color**: A coloring agent created by heating sugar in the presence of acid or alkali. It is used to give a brown color to soft drinks, sauces, and baked goods. Some types of caramel color are associated with potential carcinogenic risks when consumed in large quantities.

Recipes: Easy, Minimally Processed Meal Ideas

The following meal ideas focus on using whole, minimally processed ingredients that are both nutritious and easy to prepare. These recipes aim to provide balanced meals with plenty of vegetables, lean proteins, and whole grains, while avoiding unnecessary additives and preservatives.

1. Quinoa and Roasted Vegetable Salad

Ingredients:

- 1 cup quinoa
- 2 cups water or vegetable broth
- 1 zucchini, diced
- 1 red bell pepper, diced
- 1 red onion, diced
- 1 tablespoon olive oil
- Salt and pepper to taste
- 1/4 cup fresh parsley, chopped
- 2 tablespoons lemon juice

Instructions:

1. Rinse the quinoa under cold water. Bring water or broth to a boil in a medium pot, then add the quinoa. Reduce the heat and simmer for 15 minutes or until the quinoa is cooked and water is absorbed. Fluff with a fork.

2. Preheat the oven to 400°F (200°C). Toss the diced zucchini, bell pepper, and onion with olive oil, salt, and pepper. Spread

evenly on a baking sheet and roast for 20-25 minutes, until tender and slightly browned.

3. In a large bowl, combine the cooked quinoa with the roasted vegetables. Add parsley and lemon juice, and stir to combine. Serve warm or chilled.

This salad is packed with fiber, protein, and antioxidants, and it can be served as a side dish or a main meal. The quinoa provides a complete protein, and the vegetables are rich in vitamins and minerals.

2. Sweet Potato and Black Bean Tacos

Ingredients:

- 2 medium sweet potatoes, peeled and diced
- 1 tablespoon olive oil
- 1 can black beans, drained and rinsed
- 1 teaspoon cumin
- 1 teaspoon chili powder
- Salt and pepper to taste
- 8 small corn tortillas
- 1/4 cup fresh cilantro, chopped
- 1 lime, cut into wedges
- 1/4 cup Greek yogurt (optional, for topping)

Instructions:

1. Preheat the oven to 400°F (200°C). Toss the diced sweet potatoes with olive oil, cumin, chili powder, salt, and pepper.

Spread on a baking sheet and roast for 25-30 minutes, turning halfway, until soft and lightly browned.

2. In a small pot, heat the black beans over medium heat until warmed through.

3. Warm the tortillas in a dry skillet for 1-2 minutes on each side.

4. Assemble the tacos by placing roasted sweet potatoes and black beans on each tortilla. Top with cilantro, a squeeze of lime, and Greek yogurt if desired.

These tacos are full of plant-based protein, fiber, and vitamins. The sweet potatoes provide a healthy source of complex carbohydrates, and the black beans offer a dose of fiber and protein, making this a balanced, satisfying meal.

3. Chicken and Vegetable Stir-Fry

Ingredients:

- 1 lb chicken breast, sliced into thin strips

- 1 tablespoon olive oil

- 1 broccoli head, cut into florets

- 1 carrot, thinly sliced

- 1 bell pepper, sliced

- 2 cloves garlic, minced

- 2 tablespoons low-sodium soy sauce

- 1 tablespoon honey

- 1 teaspoon sesame oil

- Cooked brown rice, for serving

Instructions:

1. Heat olive oil in a large skillet or wok over medium-high heat. Add chicken strips and cook for 5-7 minutes, until browned and cooked through. Remove the chicken from the pan and set aside.

2. In the same pan, add the broccoli, carrot, bell pepper, and garlic. Stir-fry for 3-5 minutes, until the vegetables are tender-crisp.

3. Return the chicken to the pan. In a small bowl, whisk together soy sauce, honey, and sesame oil. Pour the sauce over the chicken and vegetables, stirring to coat.

4. Serve the stir-fry over a bed of cooked brown rice.

This stir-fry is a quick and healthy meal that combines lean protein, vegetables, and whole grains. The use of low-sodium soy sauce and honey creates a flavorful, balanced sauce without relying on excessive sugar or salt.

Resources: Recommended Books, Studies, and Organizations for Further Exploration

For those interested in diving deeper into the topics covered in this book, the following resources provide additional insights into food science, nutrition, sustainability, and the impact of processed foods.

Books

- *In Defense of Food: An Eater's Manifesto* by Michael Pollan (2008): This book explores the consequences of the industrialization of food and advocates for a return to whole foods and simple, traditional diets.

- *Food Politics: How the Food Industry Influences Nutrition and Health* by Marion Nestle (2007): This book provides an in-depth look at the ways in which the food industry shapes public health policy and influences consumer behavior.

- *The Omnivore's Dilemma: A Natural History of Four Meals* by Michael Pollan (2006): This book explores the complexities of modern food production and consumption, questioning how our food choices affect the environment and our health.

Studies

- *The Global Burden of Disease Study* (2017): This study, published by The Lancet, assesses the impact of dietary risk factors on global health outcomes, including obesity and chronic diseases.

- *The Impact of Food Processing on Nutritional Quality* by Monteiro et al. (2019): This study examines the effect of ultra-processed foods on public health and the importance of minimizing their consumption.

Organizations

- **The Food and Agriculture Organization of the United Nations (FAO)**: The FAO provides valuable information and reports on global food systems, sustainable agriculture, and nutrition.

- **The Center for Science in the Public Interest (CSPI)**: CSPI advocates for food policy reforms and provides resources on food safety, nutrition, and food labeling.

- **The Environmental Working Group (EWG)**: EWG conducts research on food safety and environmental health, providing resources on pesticides, food additives, and sustainable food choices.

www.ingramcontent.com/pod-product-compliance
Lightning Source LLC
Chambersburg PA
CBHW071726120626
46550CB00002B/394